Book T2

PUBLISHED BY THE PRESS SYNDICATE OF THE UNIVERSITY OF CAMBRIDGE
The Pitt Building, Trumpington Street, Cambridge, United Kingdom

CAMBRIDGE UNIVERSITY PRESS
The Edinburgh Building, Cambridge CB2 2RU, UK
40 West 20th Street, New York, NY 10011-4211, USA
477 Williamstown Road, Port Melbourne, VIC 3207, Australia
Ruiz de Alarcón 13, 28014 Madrid, Spain
Dock House, The Waterfront, Cape Town 8001, South Africa

http://www.cambridge.org

Printed in the United Kingdom at the University Press, Cambridge

Typeface Minion   *System* QuarkXPress®

A catalogue record for this book is available from the British Library.

ISBN 0 521 78546 4 paperback

Typesetting, technical illustrations and layout by The School Mathematics Project,
Jeff Edwards and Eikon Illustration
Other illustrations by Robert Calow and Steve Lach at Eikon Illustration
Cover image © Image Bank/Antonio Rosario
Cover design by Angela Ashton

The publishers thank the following for supplying photographs:
Pages 45 and 47, © Getty Images
Page 45, Kennel Club Picture Library
Page 86, Joyce Williams for the National Seal Santuary
Page 89, Alton Museum of History and Art, Illinois
Page 152, CJ WildBird Foods Ltd
All other photographs by Graham Portlock

The authors and publishers would like to thank Hywel Sedgwick-Jell for his help
with the production of this book

Contents

① Number bites

These are short activities to help you practise number skills.
You don't need to do them all in one go. Just use each one when it is needed.

W Whole numbers

W1 Cheque it out

On a cheque you write the amount
in words and in figures.

UPLAND BANK PLC	8 June 2002
Pay K P Stanley	£ 324 =
Three hundred and twenty four pounds	
	M L DEVLISH
	Signed M L Devlish

Write each amount below in
words.

1 £6080 **2** £5460 **3** £10 200 **4** £200 000 **5** £50 050

W2 Hidden places

Spell the name of a place by putting the numbers in order, smallest first.

1
E 69 907
I 70 883
N 71 360
L 70 660
B 69 657
R 70 099

2
D 6888
A 6686
I 6868
R 6866
M 6668
D 6688

3
R 47 007
Y 40 707
W 40 700
E 40 070
N 40 007
K 47 070
O 40 770

4
S 30 330
C 33 003
O 33 030
M 30 033
O 30 303
W 33 300

W3 Closest neighbour maze

You need sheet 218.

Start in the circle marked '300'.
Move to the neighbouring circle that has the number closest to
the number you are on. But don't visit the same circle twice!

Continue until you reach the edge of the maze again.
Record where you go and the number you finish on.

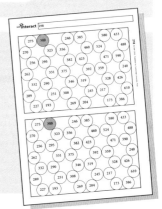

W4 Nearest hundred or thousand

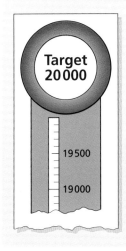

Target
20 000

19 500

19 000

1 A school is collecting stamps for charity.
   They record the number of stamps they have to the nearest 100.
   Write these numbers of stamps to the nearest 100.

   (a) 382   (b) 744   (c) 1708

   (d) 4571   (e) 12 560   (f) 18 467

2 Round these football match crowds to the nearest 1000.

Teams	Crowd
(a) Munchester Utd vs Inter Turin	42 869
(b) Chopsea vs Foulham	18 573
(c) Hadley Town vs Rawlsea Rovers	2647
(d) Blofield Tuesday vs Sorwich	12 385
(e) Loverpool vs Overton	23 096

W5 Mileometer

The mileometer on a car tells you how many miles the car has travelled since it was made.

1 One morning a car mileometer reads $\boxed{2\,4\,7\,5\,6}$. Later in the day it reads $\boxed{2\,4\,8\,5\,6}$.
   How many miles does the car go between the two readings?

2 How far does a car go between each pair of mileometer readings below?

   (a) $\boxed{6\,0\,7\,4\,8}$ $\boxed{6\,5\,7\,4\,8}$ (b) $\boxed{1\,3\,9\,2\,4}$ $\boxed{1\,3\,9\,9\,4}$

   (c) $\boxed{4\,6\,3\,0\,7}$ $\boxed{6\,6\,3\,0\,7}$ (d) $\boxed{0\,7\,0\,0\,0}$ $\boxed{1\,2\,0\,0\,0}$

3 (a) A car's mileometer reads $\boxed{1\,0\,8\,4\,7}$. What does it read after 100 more miles?

   (b) A car's mileometer reads $\boxed{7\,3\,7\,4\,4}$. What does it read after 10 000 more miles?

   (c) A car's mileometer reads $\boxed{5\,1\,4\,8\,0}$. What does it read after 3000 more miles?

W6 Sweets This activity is described in the teacher's guide.

5

W7 Pinball

Each pin has a score.

The score for this game is 70 + 35 + 125 = **230**.

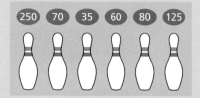

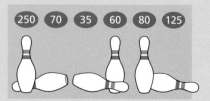

1 What is the total score when these pins have been knocked down? 250, 80, 35

2 The total score for a game is 165. Which pins were knocked down?

3 Which pins were knocked down to give each of these total scores?
   (a) 195   (b) 130   (c) 175   (d) 375   (e) 455
   (f) 445   (g) 365   (h) 490   (i) 415   (j) 480

W8 Menu This activity is described in the teacher's guide.

Menu

	Regular	Large
Hamburger	75p	95p
Sausage roll	64p	85p
Veggie-burger	69p	89p
Fries	55p	75p
Fruit pie	47p	65p
Lemonade	50p	60p
Cola	50p	60p
Tea	38p	50p

Big value hamburger meal

Large hamburger
Large fries   £1.99
Large cola or lemonade

Big value veggie-burger meal

Large veggie-burger
Large fries   £1.89
Large cola or lemonade

W9 Multiplication magic

Write down the number 12 345 679. (Don't forget to miss out the 8!)

Choose a number to multiply it by, for example 4.

```
  12345679
      × 4
  49382716
      × 9
```

Then multiply the result by 9.

Now start with 12 345 679 again and multiply by a different number.
Multiply the result by 9, as before. What happens?

W10 Remainder code

Do each division below and find the remainder. Then use the code to get a letter.

Remainder	0	1	2	3	4	5	6	7	8	9
	I	L	M	U	P	O	V	E	R	A

Rearrange the nine letters to make the name of a city.

1 $389 \div 5$ 2 $673 \div 2$ 3 $293 \div 6$

4 $5851 \div 7$ 5 $3209 \div 4$ 6 $2215 \div 8$

7 $1133 \div 9$ 8 $5286 \div 3$ 9 $6785 \div 10$

W11 Fruity sums

Work out each answer. Then use the code to find the fruit.

0	1	2	3	4	5	6	7	8	9
N	L	M	U	P	O	A	E	R	G

Example

```
   1 4 5 6
 + 2 6 7 6
   ------
   4 1 3 2
   P L U M
```

1 $37\,473 + 548\,624$ 2 5430×5 3 $33\,123 - 7028$

4 $51\,750 \div 3$ 5 $104\,523 - 5876$ 6 6901×7

7 $19\,072 \div 4$ 8 $83\,213 - 18\,796$ 9 $24\,792 \div 6$

W12 School roll

The table gives information about a school which has five year groups.

Year group	Number of girls	Number of boys
Year 7	119	105
Year 8	131	106
Year 9	118	131
Year 10	133	128
Year 11	128	126

1 How many pupils are there altogether in Year 7?

2 There are 8 classes of equal size in Year 7. How many are pupils in each class?

3 Are there more girls or more boys in the school? How many more?

4 There are nine classes in Year 9. Eight of the classes have 28 pupils each. How many pupils are there in the other class?

5 A local firm gives the school money for new computers. They give £35 for every pupil in the school. How much money is that altogether?

W13 Two sum puzzle

The six numbers 394, 259, 273, 681, 408 and 135 will make two additions with their answers.

What are the additions?

F Fractions

F1 Transport

Do each calculation.
Use the code to change the answers into letters.
Rearrange the letters to spell a means of transport.
(Some of them are unusual!)

9	14	15	24	26	28	32	34	36	40	42	45	46	50	54	70	84
L	B	I	A	N	T	O	D	S	K	R	H	C	E	Y	M	G

1 $\frac{1}{2}$ of 84

twice 12

twice 23

2 twice 20

$\frac{1}{2}$ of 28

$\frac{1}{2}$ of 30

twice 25

3 $\frac{1}{2}$ of 90

twice 21

$\frac{1}{4}$ of 200

2×18

$\frac{1}{4}$ of 128

4 $\frac{1}{2}$ of 100

$\frac{1}{4}$ of 168

twice 7

4×6

21×4

5 $\frac{1}{2}$ of 80

twice 18

$\frac{1}{2}$ of 30

$\frac{1}{4}$ of 144

6 twice 35

$\frac{1}{10}$ of 500

$\frac{1}{4}$ of 96

$\frac{1}{2}$ of 18

$\frac{1}{2}$ of 92

7 twice 17

$\frac{1}{2}$ of 84

4×21

$\frac{1}{2}$ of 18

twice 25

$\frac{1}{4}$ of 60

8 twice 20

$\frac{1}{2}$ of 100

4×9

twice 14

$\frac{1}{2}$ of 72

twice 12

9 4×8

twice 27

$\frac{1}{4}$ of 160

$\frac{1}{2}$ of 68

twice 13

$\frac{1}{8}$ of 400

10 $\frac{1}{4}$ of 36

twice 16

$\frac{1}{2}$ of 52

twice $4\frac{1}{2}$

$\frac{1}{2}$ of 64

$\frac{1}{2}$ of 48

$\frac{1}{2}$ of 28

F2 What fraction of . . .?

1 TOP is TO

2 PANE is PAN

3 STEEL is EEL

4 FRACTION is ACT

5 WEDNESDAY is WED

6 SWALLOW is ALLOW

F3 Fraction maze

You need sheet 219.

Start in the circle marked '$\frac{1}{2}$ of 20'.
Move to the neighbouring circle that has the number closest to
the number you are on. But don't visit the same circle twice!

Continue until you reach the edge of the maze again.
Record where you go and the number you finish on.

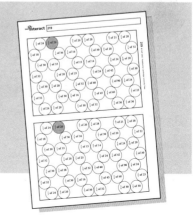

D Decimals

D1 Which fits?

Measure in **centimetres** the length and width of each coloured rectangle.
Write down the length and width of each one.

Then measure the length and width of the white hole.
Which coloured rectangle fits the hole exactly?

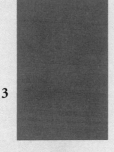

D2 Reading scales

1 What number does each arrow point to?

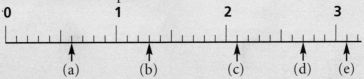

2 What number does each arrow point to?

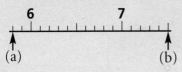 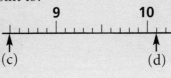 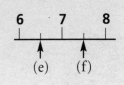

D3 'One tenth more' bingo

0.5	0.9	1	1.4
	1.8	2	2.3	2.9
3	3.5	3.9	4
	4.3	5	5.9	6

Choose six of these numbers for your Bingo card.

Your teacher calls out numbers. If you have the number
which is **1 tenth more**, you cross it off.

For example, if the teacher calls out 3.4, you can cross off 3.5.

D4 Jumping contest

Derek, Barry and Neil have a jumping contest.
Each has three jumps. Their best jump counts.

Here are the results.

Name	1st jump	2nd jump	3rd jump
Derek	4.3 m	4.6 m	4.4 m
Barry	3.9 m	4 m	4.5 m
Neil	4.1 m	3.7 m	5 m

1 The shortest jump was Neil's 2nd jump (3.7 m).
   Which was the next shortest?

2 Whose 2nd jump was his best jump?

3 Who did the best jump of all?

4 How much longer was the best jump than the next best?

D5 More or less

2.9 is 0.2 less than 3.1 .

What other statements can you make linking
the numbers on the left to the numbers on the right?

You can only use the numbers in between and 'is',
'more than' and 'less than'.

3.5		6.8
2.9	2	3.1
5.7	0.5	6.2
6	0.9	4.9
7.3	0.8 0.2	6.9
	0.4	

D6 Rodents This activity is described in the teacher's guide.

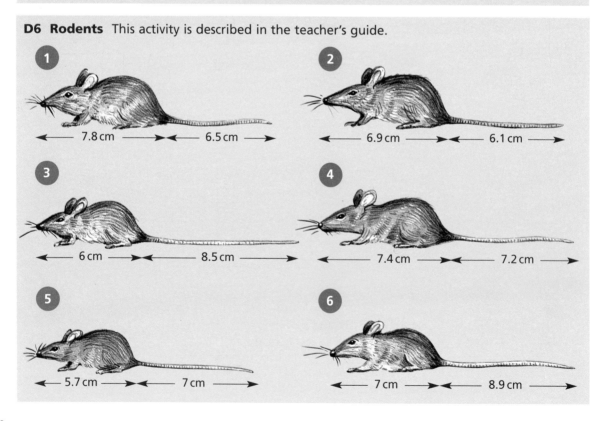

1 7.8 cm — 6.5 cm

2 6.9 cm — 6.1 cm

3 6 cm — 8.5 cm

4 7.4 cm — 7.2 cm

5 5.7 cm — 7 cm

6 7 cm — 8.9 cm

D7 Guessing games

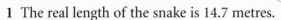

Guess the length of the snake!

20.2 m   16.3 m   13 m   11.5 m   12.9 m

1 The real length of the snake is 14.7 metres.

(a) Which guess is closest?

(b) What is the difference between the best guess and the real length?

Guess the capacity of the tank!   100 litres   152 litres   163 litres

2 The real capacity of the tank is 148.7 litres.   145 litres

(a) Which guess is closest?

(b) What is the difference between the best guess and the real capacity?

D8 Addition crosses

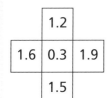

	1.2	
1.6	0.3	1.9
	1.5	

The numbers **across** make an addition sum.
$$1.6 + 0.3 = 1.9$$
The numbers **down** make an addition sum.
$$1.2 + 0.3 = 1.5$$

Draw the cross shape like this:

Copy and complete these addition crosses.

1

	0.4	
1.6	0.7	

2

	1.7	
2.3	0.8	

3

	1.3	
2.5	4	

4

	1.9	
1.4		2

5

	1.3	3.4
	2.1	

6

4.8		6
	4.7	

7

0.3	1.7	
	2.8	

8

	0.9	2.5
	5.7	

D9 Make them right

Choose numbers from this list to make the multiplications correct.

0.2   0.3   0.5   0.6   3   5

You can use each number more than once.

1   ☐ × 4 = 0.8

2   0.25 × ☐ = 0.75

3   0.3 × ☐ = 1.5

4   ☐ × 0.8 = 4

5   ☐ × 4 = 2.4

6   9 × ☐ = 1.8

D10 Packing a case

Pedro is packing a case.

He wants the total value of the things in the case to be as large as possible.

But he must not put in more than 15 kg altogether.

The table shows the things he can choose.
What should he put in the case?

	Weight	Value
Camera	2.5 kg	£145
Video recorder	4.2 kg	£276
Diamond brooch	0.1 kg	£205
Case of champagne	7.3 kg	£185
Laptop computer	5.6 kg	£912
Painting	3.7 kg	£167
Radio	3.2 kg	£ 58

D11 Shopping around

Here are the prices of some things in each of four supermarkets.

Item	Trusty	Sunbury's	Foodfair	Megavalue
500 g cornflakes	£1.20	£1.15	£0.99	£1.09
1 kg tomatoes	£1.36	£1.28	£1.40	£1.19
large white loaf	55p	59p	45p	50p
400 g digestive biscuits	77p	80p	66p	70p
80 tea bags	£1.25	£1.05	£1.30	£1.15
1 kg Granny Smith apples	£1.50	£1.40	£1.38	£1.30

1   How much does a 400 g packet of digestive biscuits cost in Foodfair?

2   (a) In which supermarket are tomatoes   (i) cheapest   (ii) dearest

    (b) What is the difference between the cheapest and dearest prices?

3   How much do these cost?

    (a) 3 large white loaves in Trusty   (b) 3 kg of apples in Sunbury's

4   I need to buy 500 g of cornflakes, a large white loaf and 400 g of digestive biscuits.
    Which would be the cheapest store to buy all these items?

② Using a spreadsheet

This work will help you use a computer spreadsheet to solve problems.

Reminders

Shorthand

* stands for ×
2*A1 means 2 × cell A1.

/ stands for ÷
A1/2 means cell A1 ÷ 2.

Formulas

Formulas in spreadsheets always start with =.

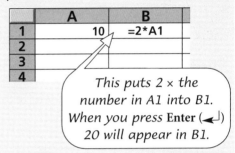

	A	B
1	10	=2*A1
2		
3		
4		

This puts 2 × the number in A1 into B1. When you press Enter (⏎) 20 will appear in B1.

Fill down

You often have to 'fill down' a formula.

First click on the cell you want to fill down.
Move the pointer to the bottom right corner.

Then drag over the cells you want to fill into.
This will select them.

When you let go of the mouse button the formulas will be dropped in the selected cells.

A	B	
	=2*A1	

A	B	
	=2*A1	

A	B	
	=2*A1	
	=2*A2	
	=2*A3	
	=2*A4	
	=2*A5	

You will not normally see the formulas.
You see the numbers that are calculated from them.
By double-clicking a cell you can see the formula in it.

Spot the formula an activity in pairs

- The second person looks away.

- The first person puts a number into a cell of the spreadsheet.
  Then they put a formula into another cell.

	A	B
1	10	
2		
3	=3*A1 + 12	
4		
5		

- When they press Enter the result of the formula will appear in the cell.

	A	B
1	10	
2		
3	42	
4		
5		

- The second person now has to find out what the formula is by trying different numbers in cell A1.

- When they think they know the formula they type it into a cell.

	A	B
1	10	
2		
3	42	
4		=4*A1 + 2
5		

- They check they are right by trying some more numbers in cell A1.

	A	B	
1	5		*Try again!*
2			
3	27		
4		22	
5			

- Take it in turns to hide and spot the formula.

Making a sequence

Type 1, 2, 3, 4, ... (up to 20) into column A of your spreadsheet.

	A	B
1	1	
2	2	
3	3	
4	4	
5	5	
6	6	

Type the formula
= A1+1
into cell B1.

	A	B
1	1	=A1+1
2	2	
3	3	
4	4	
5	5	
6	6	

Fill the formula in cell B1 down column B.

	A	B
1	1	=A1+1
2	2	
3	3	
4	4	
5	5	
6	6	

Try making some different sequences by filling down a formula in column B.

Each of the sequences in column B below
has been made by filling down a formula.

Find the formula that has been used in each case.

1

	A	B
1	1	2
2	2	4
3	3	6
4	4	8
5	5	10
6	6	12
7	7	14
8	8	16

2

	A	B
1	1	6
2	2	7
3	3	8
4	4	9
5	5	10
6	6	11
7	7	12
8	8	13

3

	A	B
1	1	3
2	2	6
3	3	9
4	4	12
5	5	15
6	6	18
7	7	21
8	8	24

4

	A	B
1	1	5
2	2	10
3	3	15
4	4	20
5	5	25
6	6	30
7	7	35
8	8	40

5

	A	B
1	1	6
2	2	11
3	3	16
4	4	21
5	5	26
6	6	31
7	7	36
8	8	41

6

	A	B
1	1	4
2	2	9
3	3	14
4	4	19
5	5	24
6	6	29
7	7	34
8	8	39

7

	A	B
1	1	3
2	2	5
3	3	7
4	4	9
5	5	11
6	6	13
7	7	15
8	8	17

8

	A	B
1	1	9
2	2	19
3	3	29
4	4	39
5	5	49
6	6	59
7	7	69
8	8	79

9

	A	B
1	1	1
2	2	4
3	3	9
4	4	16
5	5	25
6	6	36
7	7	49
8	8	64

Big, bigger, biggest

Set up your spreadsheet like this.

	A	B	
1		4	
2		5	
3			
4	**Add up to**		
5			
6	**Multiplied**		
7			
8			
9			
10			

You can type any two numbers into these two cells.

Put a formula in this cell that adds your two numbers.

Put a formula here that multiplies your two numbers together.

1 (a) Can you find two numbers that add up to 8 and when multiplied give you 12?

(b) Find two numbers that add up to 14 and give 24 when multiplied.

(c) Find two other numbers that add up to 14 but come to more than 24 when multiplied.

What is the biggest number you can get when you multiply two numbers that add up to 14?

2 Change your spreadsheet so that it adds and multiplies three numbers.

(a) Can you find three numbers that add up to 11 and come to 24 when you multiply them?

(b) What is the biggest number you can get when you multiply three numbers that add up to 11?

Furry Festivals

Furry Festivals
T-shirts   £22
Badges   £2

1 Furry Festivals sell souvenirs at festivals.
They sell T-shirts and badges.

- At one festival they sold 40 items altogether.
  They took £480 for the items.

- At another festival they also sold 40 items.
  This time they took £640 for the items.

Can you work out how many of each item they sold at each festival?

Set up a spreadsheet like the one below
to help you solve the problems.

Put here the number of T-shirts you think they sold.

Put a formula here for the cost of the T-shirts.

	A	B	C	D	E
1					
2		Cost each	Number		Total cost
3	T-shirts	22			
4	Badges	2			
5					
6	Total number			Grand total	
7					
8					
9					

Put here the total number of items they sold.

Put a formula here for the cost of all the items they sold.

Furry Festivals
T-shirts   £22
Badges   £2
Pens   £1

2 Furry Festivals add pens to the things they sell.
At the next festival they sell 60 items altogether.
They take exactly £200.
(They sell at least one of everything.)

Can you work out how many of each item they sell now?

Is there more than one answer to the problem?

③ Using rules

This is about using rules in planning.
The work will help you use rules in words or letters.

A Katy's Catering

Katy's Catering provides everything you need for a children's party.

Katy uses rules to work out

- how much food is needed for a party
- how many cups are needed

and so on.

Here are some of the rules she uses.

The number of chicken legs needed is 6 more than the number of people.

To work out the number of bottles of lemonade, divide the number of people by 2.

Take 2 paper cups for each person.

You need 3 sausage rolls for each person plus 10 extra.

A1 (a) How many chicken legs are needed for 10 people?
    (b) How many chicken legs would be needed for 20 people?

A2 How many paper cups will Katy's Catering take for
    (a) 10 people   (b) 20 people

A3 How many bottles of lemonade are needed for
    (a) 10 people   (b) 20 people

A4 How many sausage rolls are needed for
    (a) 10 people   (b) 20 people

A5 Katy's Catering takes 30 paper cups to a party.
How many people were they catering for?

18

A6 Katy uses this rule to work out the number of party poppers she needs for a party. How many party poppers are needed for a party of

You need 6 party poppers for each person plus 20 extra.

(a) 50 people

(b) 100 people

B Shorter rules

Some rules are written in a shorter way.

So this rule …

You need to take one hat for each person, and two extra.

can be written as …

number of hats = number of people + 2

B1 How many hats are needed for

(a) 18 people

(b) 30 people

B2 Here is the rule Katy's Catering uses for the number of plastic forks.

number of forks = number of people × 2

How many forks are needed for

(a) 7 people

(b) 12 people

B3 Here is a rule for the number of rubbish bags to take.

The number of rubbish bags you need is the number of people divided by 2.

It can be written much shorter.

$$number\ of\ rubbish\ bags = \frac{number\ of\ people}{2}$$

How many rubbish bags are needed for

(a) 20 people

(b) 16 people

(c) 100 people

B4 This rule says how many large bowls of trifle to take.

$$number\ of\ bowls\ of\ trifle = \frac{number\ of\ people}{6}$$

(a) How many bowls of trifle do you need for 30 people?

(b) How many bowls of trifle do you need for 90 people?

C Shorthand

Katy's Catering decide to write their formulas using a shorthand.

Here is the rule for the number of hats they need.

> number of hats = number of people + 2

They write n to stand for the *number of people*.
h to stand for the *number of hats*.

Any letters can be used as long as it is clear what they stand for.

So the hat rule becomes $h = n + 2$

Here is the rule for the number of paper cups they need.

> number of paper cups = number of people × 2

They write n to stand for the *number of people*.
c to stand for the *number of paper cups*.

You can write the cup rule as $c = n \times 2$ or, even shorter, as $c = 2n$

C1 This is the rule Katy's Catering uses for mince pies.

$$m = n + 6$$

n stands for the *number of people*.
m stands for the *number of mince pies* they take.

(a) If there are 12 people, how many mince pies will they take?

(b) If there are 8 people, how many mince pies will they take?

(c) If $n = 20$, work out what m is.

(d) If $n = 32$, what is m?

C2 This is the rule Katy's Catering uses for straws.

$$s = 3n$$

n stands for the *number of people*.
s stands for the *number of straws* they take.

(a) If there are 8 people, how many straws will they take?

(b) If there are 12 people, how many straws will they take?

(c) If $n = 10$, work out what s is.

(d) If $n = 20$, what is s?

C3 Here are three other rules Katy's Catering uses.

$$p = 4n$$ $$k = n + 5$$ $$j = \frac{n}{5}$$

Remember:
$\frac{n}{5}$ means $n \div 5$.

n stands for the *number of people.*
p stands for the *number of paper plates.*
k stands for the *number of plastic knives.*
j stands for the *number of bowls of jelly.*

40 people go to a party.

(a) How many paper plates do they need?

(b) How many plastic knives are needed?

(c) How many bowls of jelly should Katy's Catering take?

C4 Use the rules in question C3 to work these out.

(a) How many paper plates are needed at a party of 20?

(b) How many paper plates are needed at a party of 100?

(c) What is k if $n = 15$?

(d) How many bowls of jelly would 100 people need?

(e) What is j if $n = 60$?

C5 Here is the sausage roll rule as a sentence.   Here it is in shorthand.

You need 3 sausage rolls for
each person plus 10 spare.

$$r = 3n + 10$$

n stands for the *number of people.*
r stands for the *number of sausage rolls.*

(a) If n is 20, what is r?

(b) If n is 15, what is r?

(c) 50 people come to a party. How many sausage rolls are needed?

C6 Here is a rule for iced fancies.

$$f = 5n + 15$$

n stands for the *number of people.*
f stands for the *number of iced fancies* to take.

(a) Work out f when $n = 8$.

(b) What is f when $n = 20$?

(c) Copy and complete this sentence for the iced fancies.

You need ... iced fancies for each person, and another ... extra.

C7 Sometimes Katy's Catering takes crackers to parties.
Here is the rule as a sentence.

Take one cracker for each person, with 6 extra.

Which of these rules is correct for crackers?

$c = 6n$

$c = n + 6$

$c = 6n + 1$

$c = n + 1 + 6$

n stands for the *number of people*.
c stands for the *number of crackers*.

C8 This is the rule for serviettes.

Take three serviettes for each person.

Which of these rules is correct for serviettes?

$s = 3n$

$s = n + 3$

$s = 3 + n$

$s = \dfrac{n}{3}$

n stands for the *number of people*.
s stands for the *number of serviettes*.

*C9 This is the rule for loaves of bread.

You need half a loaf of bread for each person.

Which of these rules is correct for loaves?

$l = \dfrac{1}{2} + n$

$l = n + \dfrac{1}{2}$

$l = 2n$

$l = \dfrac{n}{2}$

n stands for the *number of people*.
l stands for the *number of loaves*.

*C10 Which of these rules is short for

You need twice as many bread rolls as people, plus 6 more.

$r = n + 6$

$r = 6n + 2$

$r = 2n + 6n$

$r = 2n + 6$

$r = 2 + 6$

n stands for the *number of people*.
r stands for the *number of rolls*.

D Working out formulas

D1 Expocamp supply equipment to expeditions.
They use rules to decide how many of each thing to take.
In each rule, *n* stands for the *number of people* on the expedition.

$$t = \frac{n}{4}$$

t is the number of tents.

$$p = 12n + 2$$

p is the number of packets of porridge.

$$s = 4n - 2$$

s is the number of cans of sardines.

$$r = n + 6$$

r is the number of climbing ropes.

(a) Expocamp organise an expedition for 20 people.

    (i)   Work out how many tents they take.

    (ii)   How many packets of porridge do they take?

    (iii) How many cans of sardines?

    (iv) How many climbing ropes will they take?

(b) Work out the numbers of each item for
an expedition of 12 people.

(c) Work out the numbers for 40 people.

D2 Ms Prism orders stock for the maths department.
Naturally she uses rules to decide what to buy.

Here are some of the rules she uses.
(*c* stands for the *number of children* in that year.)

$$t = \frac{c}{4}$$

t is the number of textbooks.

$$m = c - 20$$

m is the number of mirrors.

$$u = 4c + 50$$

u is the number of multilink cubes.

$$w = \frac{c}{10} - 10$$

w is the number of wallcharts.

Work out how many of each item she has to buy for

(a) 200 children   (b) 100 children   (c) 1000 children

D3 Ms Prism buys two pencils for each child, plus an extra 50.
Write this as a shorthand rule.

D4 Noddy's Nursery Needs sell equipment for children's nurseries.

This is part of a leaflet that Noddy's give to people who are thinking of starting a nursery.

Noddy's Nursery Nuggets

Think carefully before you begin. You will need to buy a great deal of equipment.

Suppose you are starting up a nursery for n children. Here are some handy rules about what you will need.

You will need to buy $n + 4$ chairs.

We strongly recommend buying $2n + 6$ bibs.

We suggest that $5n - 10$ building blocks will be most useful.

At least $\frac{n}{2} + 1$ small tables will be needed.

You will find $2n - 2$ of our tidyboxes invaluable.

Potties are always a problem.

Work out how many of each item you would need to start up a nursery for

(a) 20 children   (b) 6 children   (c) 10 children

D5 Use the rule $n = 3m + 5$ to work out n when

(a) $m = 5$ (b) $m = 10$ (c) $m = 1$ (d) $m = 0$

D6 Use the rule $y = 2x - 6$ to work out y when

(a) $x = 20$ (b) $x = 5$ (c) $x = 3$ (d) $x = 100$

D7 Use the rule $f = \frac{x}{4} - 6$ to work out f when

(a) $x = 40$ (b) $x = 100$ (c) $x = 50$ (d) $x = 200$

*D8 Here is a rule. Work out r when $r = 5(g + h)$

(a) $g = 3$ and $h = 5$ (b) $g = 4$ and $h = 1$ (c) $g = 10$ and $h = 2$

*D9 Here is a rule. Work out z when $z = 2(y - x)$

(a) $y = 5$ and $x = 1$ (b) $y = 4$ and $x = 0$ (c) $y = 100$ and $x = 10$

What progress have you made?

Statement   **Evidence**

I can use a simple rule in words.

1 When I take my class on a picnic, I take 4
sandwiches for each child, plus 6 extra.
How many sandwiches will I take for 12
children?

2 To work out how many bottles of cola to take,
I divide the number of children by 2, and then
take off 4.
How many bottles will I need for 30 children?

I can use shorter rules.

3 I always take tablecloths on picnics.
I work out how many to take using the rule

$$number\ of\ tablecloths = \frac{number\ of\ children}{4}$$

How many tablecloths will I take for 20
children?

I can use simple rules written
with letters.

4 This is the rule I use for working out how
many paper hankies to take on picnics.
$$h = 2c$$
h is the *number of hankies.*
c is the *number of children.*

How many hankies will I take for 10 children?

5 This is the rule I use for sticking plasters.
$$p = c + 5$$
p is the *number of plasters.*
c is the *number of children.*

How many plasters do I take for 8 children?

I can use more complicated rules
written with letters.

6 Here is a rule. $h = 4c + 3$
Work out *h* when

(a) $c = 10$ (b) $c = 2$ (c) $c = 100$

7 Use the rule $y = 4x - 1$ to work out *y* when

(a) $x = 3$ (b) $x = 10$ (c) $x = 100$

④ Repeats

This work will remind you about coordinates and patterns.

T

Here is a pattern using a simple ghostly shape.

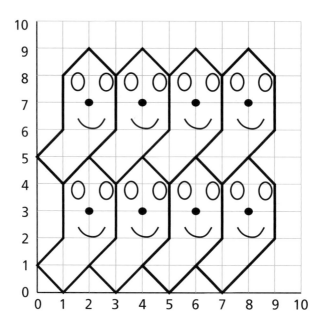

1 The picture on the right shows another ghostly pattern.
The eyes are labelled A to N.

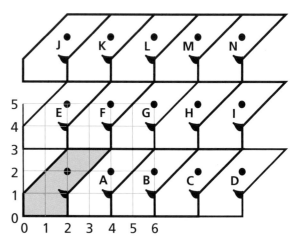

(a) What are the coordinates of the eye on the coloured shape?

(b) Copy and complete this sentence.
To move from eye A to eye B you add ... to the ... coordinate.

(c) Find the coordinates of
(i) C   (ii) D

(d) Copy and complete this sentence.
To move from eye A to eye F you add ... to the ... coordinate.

(e) Find the coordinates of
(i) J   (ii) L

(f) Find the coordinates of
(i) I   (ii) M   (iii) N

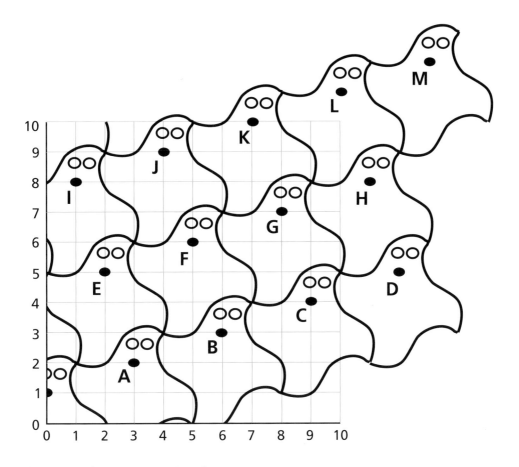

2 This is another pattern using ghosts.

   (a) Write down the coordinates of mouths A and B.

   (b) (i) What do you have to add to the *x*-coordinate
          to move from mouth A to mouth B?

      (ii) What do you have to add to the *y*-coordinate
           to move from A to B?

   (c) Do you add the same numbers moving from mouth B to C?

   (d) Find the coordinates of mouth D.

3 (a) Copy and complete this sentence for going from mouth A to F.
      You add ... to the x-coordinate and add ... to the y-coordinate.

   (b) Find the coordinates of mouth H.

   (c) Find the coordinates of mouth M.

4 The coordinates of the tip of ghost E's tail are (1, 2).

   (a) Find the coordinates of ghost M's tail.

   (b) Find the coordinates of ghost A's tail.

⑤ Multiples and factors

This work will help you recognise multiples and factors of numbers.

A Multiples

Multiples of 3 are numbers which are in the 3 times table:

> 3, 6, 9, 12, 15 and so on.

Multiples of 3 can be divided exactly by 3. For example, 60 is a multiple of 3.

Multiples on the hundred square

On sheet 134, shade or colour all the multiples of 3.

1	2	3	4	5	6	7	8	9	10
11	12	13	14	15	16	17	18	19	20
21	22	23	24	25	26	27	28	29	30

On a new hundred square, shade or colour all the multiples of 4.

1	2	3	4	5	6	7	8	9	10
11	12	13	14	15	16	17	18	19	20
21	22	23	24	25	26	27	28	29	30

Do the same for multiples of other numbers.
Describe the patterns you get.

B Multiplying

B1 Here are 5 rows of windows.
There are 4 windows in each row.

Copy and complete: **5 × 4 = ...**

B2 Write a multiplication for each of these.

(a)

(b)

(c)

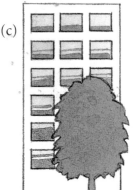

B3 Write a multiplication for each of these.

(a)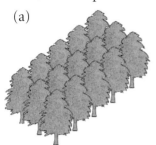

(b)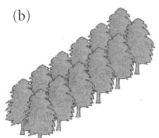

(c)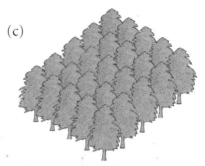

3 rows of trees,
5 in each row

2 rows of trees,
6 in each row

5 rows of trees,
5 in each row

B4 Work these out.

(a) 2 × 7   (b) 3 × 5   (c) 5 × 2   (d) 6 × 5   (e) 5 × 7

B5 What are the missing numbers?

(a) 2 × **?** = 10   (b) 3 × **?** = 9   (c) **?** × 2 = 16   (d) 5 × **?** = 30   (e) 4 × **?** = 12

B6 Are these numbers multiples of 3? Write 'yes' or 'no' for each one.

(a) 12   (b) 8   (c) 20   (d) 27

B7 (a) Is 13 a multiple of 3?   (b) Is 14 a multiple of 4?

(c) Is 15 a multiple of 5?   (d) Is 16 a multiple of 6?

B8 Which numbers in the box are multiples of 3?

3 6 9 10 12 14 15 16 21 30 31

B9 Which numbers in this box are multiples of 5?

5 8 10 12 15 20 24 25 45 48 50 55

B10 (a) Which numbers in this box are multiples of 2?

6 9 10 14 15 17 18 20 21 26 30

(b) What do we call numbers which are multiples of 2?

B11 (a) Write down any five multiples of 10.

(b) Pick out the multiples of 10 from the box.

70 90 100 103 110 148 220

(c) How can you tell by looking at a number that it is a multiple of 10?

B12 Here is a list of numbers.

20, 21, 22, 23, 24, 25, 26, 27, 28, 29, 30

Which of these numbers are multiples of

(a) 2   (b) 3   (c) 4   (d) 5   (e) 6

(f) 7   (g) 8   (h) 9   (i) 10   (j) 11

C Missing multiples

C1 How many windows are there in each picture?

(a)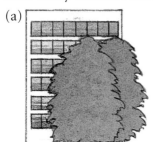

(b)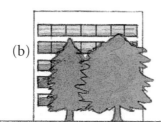

(c)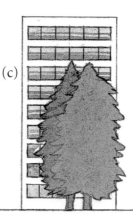

C2 There are 24 windows in this block.
How many are there in each row?

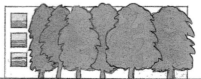

C3 How many windows are there in each row?

(a)

20 windows altogether

(b)

21 windows altogether

(c)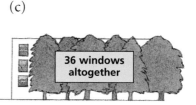

36 windows altogether

C4 Work these out.

(a) 4 × 8   (b) 3 × 9   (c) 7 × 5   (d) 6 × 6   (e) 7 × 4

C5 What are the missing numbers?

(a) 6 × **?** = 24   (b) 4 × **?** = 36   (c) **?** × 5 = 45   (d) **?** × 7 = 21

C6 What are the missing numbers?

(a) 6 × **?** = 30   (b) 7 × **?** = 42   (c) **?** × 6 = 54   (d) **?** × 7 = 28

C7 Find as many different ways as you can to complete   ... × ... = 12

C8 Find as many different ways as you can to complete   ... × ... = 20

C9 Here is a list of numbers.   6, 8, 12, 15, 18, 20, 24, 30

(a) Which numbers are multiples of 2?   (b) Which are multiples of 3?

(c) Which are multiples of 4?   (d) Which are multiples of 5?

C10 Pioneer Buses decide to make all their fares multiples of 5p.

Which of these fares
will have to be changed?

15p 20p 33p 45p 59p 66p 75p 90p

C11 Find as many ways as you can to complete   ... × ... = 30

C12 What different ways can you complete these
- (a) 6 is a multiple of ...
- (b) 8 is a multiple of ...
- (c) 12 is a multiple of ...
- (d) 20 is a multiple of ...

Nasty multiples game
A game for 2 players

The board and rules for this game are on sheet 220.

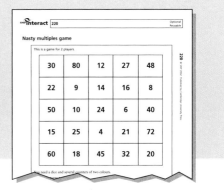

D Factors

12 can be divided exactly by 3.
We say 3 is a **factor** of 12.

Other factors of 12 are 1, 2, 4, 6 and 12.

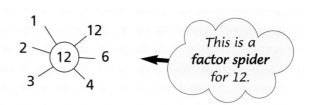

This is a factor spider for 12.

D1 Copy and complete these factor spiders.

(a) 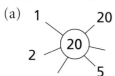 (b) 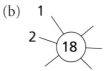 (c) 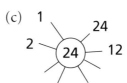 (d) ⌀16

D2 One factor of 30 is missing from this list. What is it?

    1, 2, 3, 5, 10, 15, 30

D3 Two factors of 40 are missing from this list. What are they?

    1, 2, 4, 5, 8, 10

D4 Which word, 'multiple' or 'factor', goes in each of these?
- (a) 6 is a .... of 2
- (b) 6 is a .... of 30
- (c) 10 is a .... of 60
- (d) 8 is a .... of 32
- (e) 8 is a .... of 4
- (f) 36 is a .... of 9
- (g) 15 is a .... of 30
- (h) (Careful!) 6 is a .... of 6

E Factor codes

Boris wants to send this message.   **SEND EXTRA SUPPLIES NOW**

He counts the letters and finds there are 20.
He thinks of two numbers which multiply to give 20.   (4 × 5) ○ ○

He draws a grid 4 units by 5.
He writes the message in it, like this.

S	E	N	D	E
X	T	R	A	S
U	P	P	L	I
E	S	N	O	W

Now he reads down each column
and writes the letters like this.   **SXUEETPSNRPNDALOESIW**

He sends the message to Suki.

Suki gets the coded message   **SXUEETPSNRPNDALOESIW.**
She counts the letters: 20.
She doesn't know what grid Boris drew.

> I could try 5 × 4, or 4 × 5, or 2 × 10, or 10 × 2.

She tries a 2 × 10 grid first.

She writes the message **downwards** ...
... and reads it **across**:

SUEPNPDLEIXETSRNAOSW

S	U	E	P	N	P	D	L	E	I
X	E	T	S	R	N	A	O	S	W

It doesn't make sense!

Next she tries a 4 × 5 grid.

This time the message makes sense.

S	E	N	D	E
X	T	R	A	S
U	P	P	L	I
E	S	N	O	W

E1 Here is a coded message.   **P R Y T E A T A T N**

There are 10 letters.

The grid could be 2 × 5   or 5 × 2

Try both and see which will decode the message.

E2 Decode each of these messages.
You may need to try different grids.

(a) MTAOEMTUEEFR

(b) TETRAEAVIIEGNSHLAT

(c) LBSEESAHHVIEENLBDTABEGUR

(d) WETIOIYSNNLOIEILUDMGSOEAHEUCTT

E3 Make up a message.
Use a grid to write it in code.

Then give your coded message to someone.
See if they can decode it.

Multiples and factors maze

You need sheet 221 or 176.
The rules for the maze are printed on the sheet.

What progress have you made?

Statement	Evidence
I know what 'multiple' means.	1 Write down three multiples of 4.
I can recognise multiples of a number.	2 Which of these numbers are multiples of 6? 6, 10, 12, 15, 18, 20, 24
I know what 'factor' means.	3 Write down all the factors of 20.
I can recognise factors of a number.	4 Which of these numbers are factors of 30? 1, 2, 3, 4, 5, 6, 10, 12, 15, 30
I don't confuse factor and multiple.	5 Which word, 'factor' or 'multiple', goes in these? (a) 15 is a .... of 5 (b) 4 is a .... of 32 (c) 12 is a .... of 24 (d) 36 is a .... of 12

⑥ Puzzle it out

These activities will help you practise number skills.

A Road map

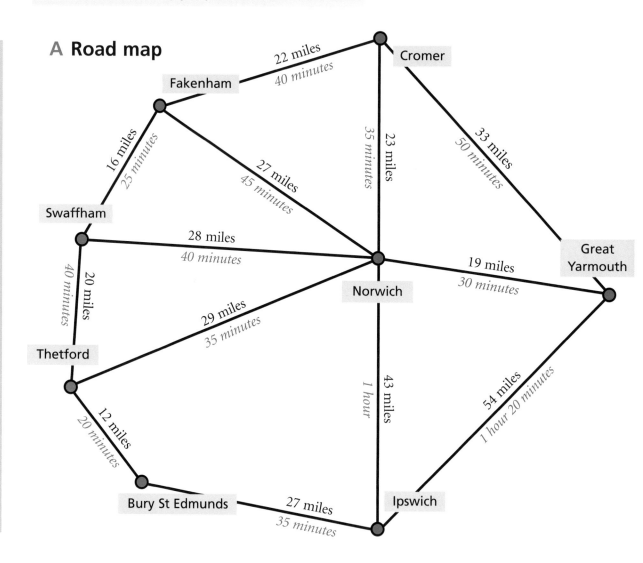

B Countup

73

4, 1, 5, 4, 7, 10

123

2, 2, 3, 5, 8, 9

169

2, 3, 5, 6, 10, 25

⑦ Two decimal places

This work will help you
- ◆ understand numbers with two decimal places
- ◆ put decimals in order

A Tenths review

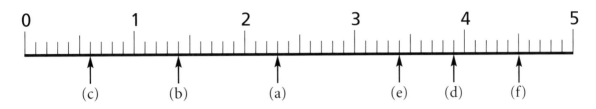

A1 What number does each arrow above point to?

A2 What number (as a decimal) is halfway between 2 and 3?

A3 Find 2.9 on the number line. What is 2.9 + 0.1?

A4 Find 1 on the number line. What is 1 – 0.1?

A5 Write these numbers in order, smallest first. 3, 4.6, 2.8, 5, 0.9

A6 Katy did 1.3 + 2 like this.   What should the answer be?

A7 Work these out.
(a) 2.3 + 1.4   (b) 3.5 + 4.7   (c) 5 + 2.6   (d) 3.6 + 4

A8 Work these out.
(a) 5.7 – 2.3   (b) 8.5 – 2.9   (c) 7 – 1.3   (d) 6.6 – 4

A9 Do these in your head.
(a) Add 1 to 23.6   (b) Add 10 to 23.6   (c) Add 0.1 to 23.6
(d) 145.6 + 10   (e) 145.6 + 0.1   (f) 10 – 0.1

A10 What do these figures stand for?
(a) The 3 in 234.1   (b) The 7 in 431.7   (c) The 4 in 530.4

B Tenths and hundredths

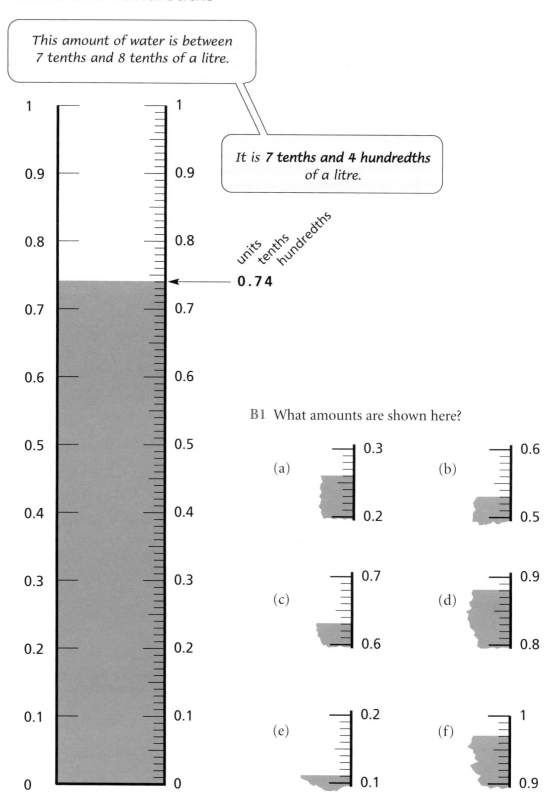

This amount of water is between 7 tenths and 8 tenths of a litre.

It is **7 tenths and 4 hundredths** of a litre.

units tenths hundredths

0.74

B1 What amounts are shown here?

(a)

(b)

(c)

(d)

(e)

(f)

B2 What number does each arrow point to?

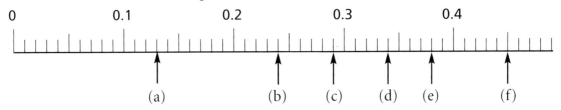

B3 (a) What number is halfway between 0.2 and 0.3?
     (The diagram above may help.)

   (b) What number is halfway between 0.1 and 0.2?

   (c) What number is halfway between 0.3 and 0.4?

   (d) What number is halfway between 0.6 and 0.7?

B4 What is the length of each of these pins, in cm?

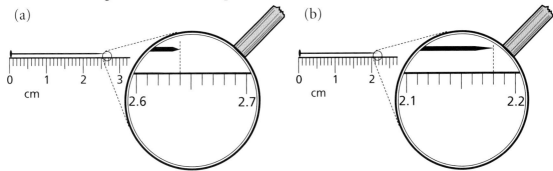

B5 What number does each arrow point to?

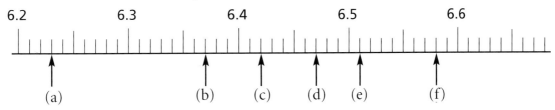

B6 (a) What number is halfway between 6.2 and 6.3?
     (The diagram above may help.)

   (b) What number is halfway between 6.4 and 6.5?

   (c) What number is halfway between 6.5 and 6.6?

   (d) What number is halfway between 6.7 and 6.8?

B7 What numbers are
   marked with arrows?

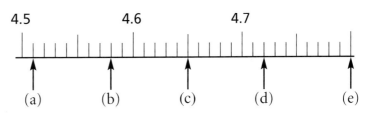

C Less than 0.1

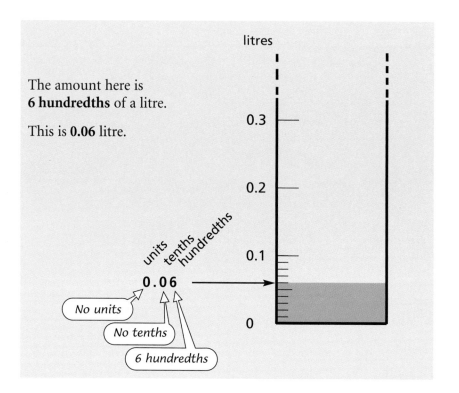

The amount here is **6 hundredths** of a litre.

This is **0.06** litre.

units tenths hundredths

0.06

No units

No tenths

6 hundredths

C1 What amounts are shown here?

(a) 0.1   (b) 0.1   (c) 0.1   (d) 0.1

C2 What number does each arrow point to?

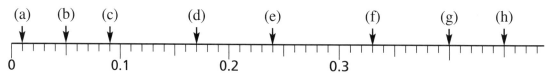

(a) (b) (c)   (d)   (e)   (f)   (g) (h)

0   0.1   0.2   0.3

C3 This question is on sheet 223.

C4 Write these as decimals.

(a) five hundredths   (b) 9 hundredths   (c) one hundredth

D 0.3 and 0.30 are equal

The top number line is divided into **tenths**.
3 tenths, or **0.3**, is marked with an arrow.

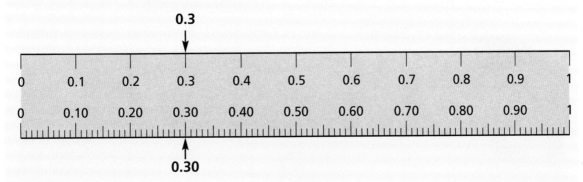

The bottom number line is divided into **hundredths**.
30 hundredths, or **0.30**, is marked with an arrow.

0.3 and **0.30** are equal.

D1 Use the number line above to help you
    (a) write another decimal equal to 0.7
    (b) write another decimal equal to 0.1
    (c) write another decimal equal to 0.80
    (d) write another decimal equal to 0.40

D2 (a) What number is halfway between 0.7 and 0.8?
    (b) What number is halfway between 0.9 and 1?
    (c) What number is halfway between 0 and 0.1?

D3 Find where these numbers are on the number lines above.
    0.05, 0.5, 0.3
  Write them in order of size, smallest first.

D4 Write each list of numbers in order of size, smallest first.
  Use the number lines to help you.
    (a) 0.9, 0.2, 0.65   (b) 0.4, 0.15, 0.08   (c) 0.62, 0.3, 0.07, 0.1

D5 Write these lists in order, smallest first,
  without looking at the number lines.
    (a) 0.6, 0.03, 0.25, 0.61   (b) 0.1, 0.06, 0.3, 0.17   (c) 0.55, 0.6, 0.08, 0.5

E Decimals of a metre

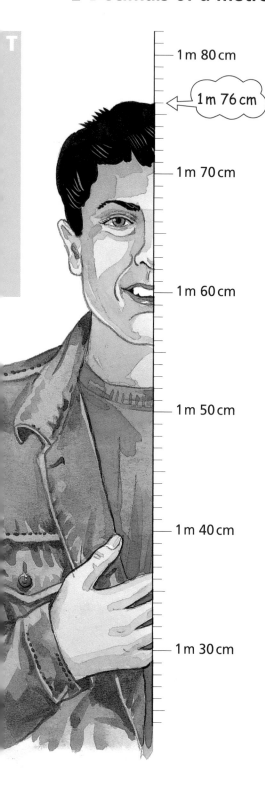

1 m 80 cm

1 m 76 cm

1 m 70 cm

1 m 60 cm

1 m 50 cm

1 m 40 cm

1 m 30 cm

A metre is divided into 100 **centimetres**.
Each centimetre is **one hundredth** of a metre.

A height of 1 metre and 76 centimetres can be written as
1.76 metres ('one point seven six metres').

- One way to write **1 metre and 30 cm**
  is 1.30 metres.

  Enter 1.30 into a calculator and press $\boxed{=}$.
  It says 1.3. Why?

- Which is greater, 1.8 metres or 1.38 metres? Why?

- How do you write in decimals **1 metre and 5 cm**?

E1 Write these in metres using decimals.

(a) 3 metres and 25 centimetres

(b) 4 metres and 68 centimetres

E2 Write 139 cm in   (a) metres and cm   (b) metres

E3 Write these in metres.

(a) 1 metre and 70 cm   (b) 1 metre and 7 cm

(c) 2 metres and 45 cm   (d) 3 metres and 6 cm

E4 (a) Hitesh is 147 cm tall. Write his height in metres.

(b) Sadia is 107 cm tall. Write her height in metres.

(c) Gail is 160 cm tall. Write her height in metres.

E5 Copy and complete this table.
It shows the heights of three people.

Name	Height in cm	Height in metres and cm	Height in metres
Alan	167 cm		
Kira		1 m 8 cm	
Greg			1.4 m

E6 Here are the lengths of four snakes.

  Cobra 3.05 m   Python 3.17 m   Boa constrictor 3.3 m   Mamba 3.1 m

(a) Which is the shortest snake?

(b) Which is the longest?

E7 Put these lengths in order, shortest first.

  1.4 m   1.18 m   1.66 m

E8 Put these lengths in order, shortest first.

  0.7 m   0.53 m   0.09 m

E9 This is part of a measuring tape.
What numbers do the arrows point to?

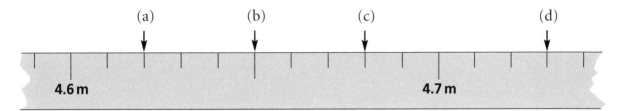

E10 What numbers do these arrows point to?

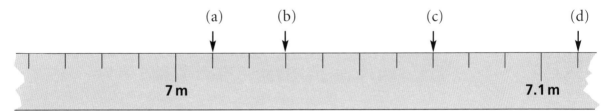

E11 This question is on sheet 224.

Height and armspan

Measure the height and armspan of some people.
Record the measurements in **metres**.

Look at the heights and armspans you recorded.
Is one always greater than the other?

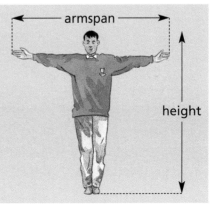

F Ordering decimals

F1 Which of these numbers are between 2.3 and 2.5?

2.35   2.55   2.4
2.25   2.04

F2 Which of these numbers are between 5 and 5.2?

0.51   5.17   5.03
5.1   5.21

F3 Write each list in order of size, smallest first.

(a) 5.32, 5.84, 6, 5.09, 5.76

(b) 3.2, 3.07, 2.8, 3.19, 3.5

(c) 0.6, 0.15, 1.07, 1.1, 0.45

F4 Spell a word by arranging the numbers in order, smallest first.

A	D	O	L	I	S	H	Y
0.36	0.2	0.08	0.1	0.16	0.41	0.03	0.4

F5 Spell another word by arranging these numbers in order.

J	A	B	O	N	E	L	E	Y
1.57	1.96	2.05	1.6	1.3	1.14	2.26	2.4	1.8

F6 Spell a word by arranging these numbers in order.

S	O	R	U	D	A	N	I
0.61	0.6	1.5	1.07	0.08	0.9	0.54	0.2

F7 This table gives the results of a long jump contest.

Name	1st jump	2nd jump	3rd jump
Singh	3.92 m	4.36 m	4.5 m
Barker	4.2 m	4.13 m	3.87 m
Church	4.18 m	4.4 m	4.12 m

(a) How long was Barker's longest jump?

(b) Whose 1st jump was longest?

(c) Whose 2nd jump was longest?

(d) What was the length of the longest jump of all?

G Place value

thousands hundreds tens units tenths hundredths

$$2\ 7\ 1\ 3\ .\ 5\ 8$$

Do not use a calculator in this section.

G1 The figure 5 in the number 2713.58 above stands for 5 tenths, or 0.5
What do these figures stand for?

(a) The 7   (b) The 2   (c) The 3   (d) The 1   (e) The 8

G2 Do these in your head.

(a) Add a ten to 2713.58   (b) Add a thousand to 2713.58

(c) Add a tenth to 2713.58   (d) Add a unit to 2713.58

(e) Add a hundredth to 2713.58   (f) Add a hundred to 2713.58

G3 Do these in your head.

(a) Add 100 to 1426.35   (b) Add 1 to 1426.35

(c) Add 0.1 to 1426.35   (d) Add 1000 to 1426.35

(e) Add 10 to 1426.35   (f) Add 0.01 to 1426.35

G4 What do these figures stand for?

(a) The 7 in 16.72   (b) The 3 in 4.03   (c) The 6 in 2364.81

(d) The 8 in 735.82   (e) The 4 in 320.04   (f) The 5 in 13.58

G5 Jason started with the number **463.86**
He subtracted something and ended up with **403.86**
What number did he subtract?

G6 Pam started with **248.71**
She subtracted a number and ended up with **248.01**
What number did she subtract?

G7 Grant started with **54.73**
He subtracted a number and ended up with **54.7**
What did he subtract?

What progress have you made?

Statement

Evidence

I can read scales to two decimal places.

1 What number does each arrow point to?

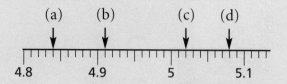

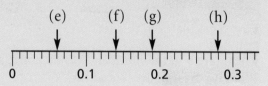

I can use decimals of a metre.

2 (a) Write 1 metre 45 centimetres in metres.

(b) Write 2 metres 3 centimetres in metres.

(c) Write 3.2 metres in metres and centimetres.

I can order numbers with up to two decimal places.

3 Write each list of numbers in order of size, smallest first.

(a) 7.3, 6.68, 6.9, 7.04, 6.7

(b) 3.24, 0.07, 0.1, 4, 1.56

I know what the figures in decimal numbers stand for.

4 What do these figures stand for?

(a) The 8 in 12.83

(b) The 4 in 23.04

(c) The 7 in 100.67

Spot the ball

This work will help you to use a grid with negative *x*-coordinates.

1 Can you guess where the tennis ball should be in this picture?

2 This dog is playing 'Flyball'.
   Where do you think the ball is that she is trying to catch?

Review 1

1 The Six Valleys Railway organise special Santa trains at Christmas.
They give mince pies and other goodies to passengers on the train.

They use rules to work out what to take on the train.

> number of mince pies = number of passengers − 20

> number of crackers = number of passengers + 10

> number of streamers = $\dfrac{\text{number of passengers}}{2}$

> number of party poppers = number of passengers × 2

On December 20th there are 120 passengers on the train.

(a) How many mince pies will they take on the train?

(b) How many crackers will they take?

(c) How many streamers will be taken?

(d) How many party poppers will they take?

On December 21st there are 200 passengers.

(e) Work out how many of each item they will take.

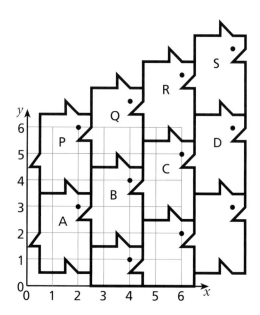

2 Use the rule $a = 2b + 1$ to work out a when

(a) $b = 3$ (b) $b = 5$ (c) $b = 10$ (d) $b = 100$

3 Use the rule $p = \dfrac{q}{4} + 2$ to work out p when

(a) $q = 20$ (b) $q = 40$ (c) $q = 24$ (d) $q = 100$

4 Look at this pattern of beasts.

(a) What are the coordinates of the eye
of beast A?

(b) What are the coordinates of the eye of B?

(c) Work out what the coordinates
of the eye of beast D must be.

(d) What are the coordinates of beast P's eye?

(e) Work out the coordinates of the eye of

   (i) beast Q   (ii) beast R

   (iii) beast S

5 Which numbers in the box are multiples of 4?

2 4 6 8 12 14 18 20 28 34 38 42

6 Which numbers in the box are factors of 60?

1 2 3 4 5 6 7 8 9 10 12 15 18 20

7 Copy and complete these using the word 'factor' or 'multiple' in each.

(a) 2 is a .... of 12

(b) 8 is a .... of 4

(c) 24 is a .... of 2

(d) 8 is a .... of 64

8 A school caretaker has to arrange 120 chairs in the hall. He could arrange them all in a rectangle with 10 rows and 12 chairs in each row.

Work out three other ways that he could arrange all the chairs in a rectangle.

9 What number does each arrow point to?

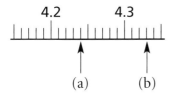
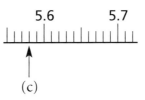
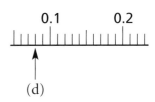

10 Gareth is 167 cm tall. Rachel is 1 m 8 cm tall. Who is taller? By how many centimetres?

11 Write these lengths in order, shortest first.

1.5 m   128 cm   1.09 m   1 m 1 cm   1.10 m

12 Do these in your head.

(a) Add a hundredth to 234.56

(b) Add $\frac{1}{10}$ to 234.56

(c) Add a ten to 234.56

(d) Add a hundred to 234.56

13 Some friends are playing 'Spot the ball'.

(a) Write down the coordinates of each guess.

(b) The ball is actually at (⁻3, 4). Whose guess was nearest?

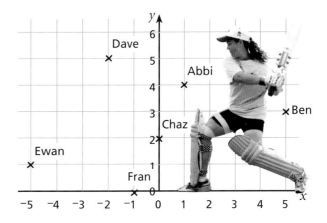

⑨ Hexathings

This work will help you

◆ practise using numbers in formulas

◆ use the correct formula for a situation

A Creepy crawlies

Here are three caterpillars.

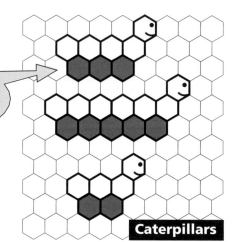

This caterpillar is made from 3 red hexagons and 5 white hexagons.

Caterpillars

• Look at the caterpillar with 5 red hexagons.
How many whites does it have?

• Look at the caterpillar with 2 red hexagons.
How many whites are there?

Can you tell how many whites a caterpillar with 7 red hexagons will have?

A1 (a) On sheet 226, draw a caterpillar with 4 red hexagons.

   (b) How many white hexagons are there in your caterpillar?

A2 (a) Draw a caterpillar with 6 red hexagons.

   (b) How many whites does it have?

A3 Copy and complete this table for the caterpillars.

Number of red hexagons	2	3	4	5	6
Number of white hexagons					

A4 (a) How many white hexagons would a caterpillar with 10 reds have?

   (b) How many whites would a caterpillar with 100 reds have?

A5 Which of these rules is correct for the number of white hexagons
when you know the number of reds?

number of white hexagons = + 2

number of white hexagons = number of red hexagons + 2

number of white hexagons = number of red hexagons – 2

number of white hexagons = number of red hexagons + 1

A6 These patterns are called moths.

Look at the moth with
3 red hexagons.
How many white hexagons does it have?

A7 (a) How many white hexagons does the
moth with 5 red hexagons have?

(b) How many white hexagons are there
in the moth with 6 red hexagons?

A8 (a) On sheet 226 draw a moth
with 4 red hexagons.

(b) How many white hexagons
does your moth have?

A9 Draw a moth with 2 red hexagons.

A10 Without drawing, think of a moth
with 10 red hexagons.
How many white hexagons would it have?

A11 Think of a moth with 100 red hexagons.
How many whites would it have?

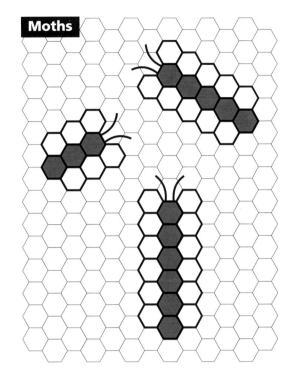

Moths

A12 Copy and complete this table for the moths.

Number of red hexagons	2	3	4	5	6		10	100
Number of white hexagons								

A13 Which of these rules is correct for the number of white
hexagons when you know the number of reds a moth has?

number of white hexagons = number of red hexagons + 2

number of red hexagons = number of white hexagons x 2

number of red hexagons = ÷ 2

number of white hexagons = number of red hexagons x 2

A14 Suppose you have a moth with 8 red hexagons.

(a) Use your rule to work out how many whites it will have.

(b) Draw the moth to check your rule works.

A15 These are earwigs!

(a) Draw earwigs with 1, 2 and 5 red hexagons.

(b) Without drawing, how many white hexagons would a 10 red hexagon earwig have?

(c) How many whites would a 100 red hexagon earwig have?

(d) Copy and complete the table below for the earwig family.

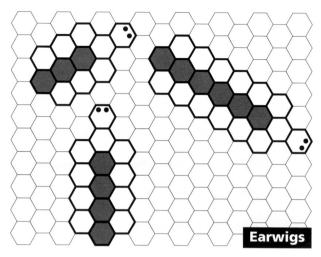

Earwigs

Number of red hexagons	1	2	3	4	5	6	10	100
Number of white hexagons								

A16 Which of these rules is correct for the earwigs?
(There might be more than one!)

> number of white hexagons = number of red hexagons + 2

> number of white hexagons = (number of red hexagons × 2) + 2

> number of white hexagons = (number of red hexagons + 1) × 2

> number of white hexagons = number of red hexagons × 2

A17 (a) Use your rule to work out how many white hexagons an earwig with 8 reds would have.

(b) Draw the earwig with 8 reds. Check your answer to part (a) is correct.

Invent your own family

Look at this rule.

number of white hexagons = number of red hexagons + 3

• Copy and complete this table for the rule.

Number of red hexagons	2	3	4	5	6
Number of white hexagons	5				

• Draw a family of hexathings that fits the rule.
Give a name to your family.

B Shorter rules

number of white hexagons = (number of red hexagons × 2) + 3

You don't need to write 'number of white hexagons'
and 'number of red hexagons' all the time.
You can use shorthand.

You can write *w* to stand for the *number of white hexagons*
and *r* to stand for the *number of red hexagons*.

So the rule above becomes $w = (r \times 2) + 3$,
or even shorter $w = 2r + 3$.

We write r × 2 or 2 × r as 2r.

B1 Match each of these rules with one of the
shorthand rules in the box on the right.
Beware! Two rules in the box are not needed.

(a) *number of white hexagons = (number of red hexagons × 3) + 2*

(b) *number of white hexagons = (number of red hexagons × 2) + 2*

(c) *number of white hexagons = (number of red hexagons × 2) + 3*

(d) *number of white hexagons = number of red hexagons + 3*

(e) *number of white hexagons = number of red hexagons × 3*

$w = 2r + 3$
$w = 3r$
$w = 2r + 2$
$w = 3r + 3$
$w = r + 3$
$w = 3r + 2$
$w = 2r - 2$

B2 Write each of these rules in shorthand.

(a) *number of white hexagons = (number of red hexagons × 4) + 2*

(b) *number of white hexagons = number of red hexagons + 2*

(c) *number of white hexagons = number of red hexagons × 5*

(d) *number of white hexagons = (number of red hexagons × 3) + 5*

B3 Here is a shorthand rule.

$w = 3r + 4$

w stands for the *number of white hexagons*.
r stands for the *number of red hexagons*.

(a) If the number of red hexagons is 6,
what is the number of white hexagons?

(b) When $r = 10$, work out what *w* is.

(c) When $r = 20$, what is *w*?

(d) If the number of **white** hexagons is 10,
what is the number of **red** hexagons?

C Chronic creatures

C1 These are snails.

(a) How many white hexagons does a snail with 4 reds have? Do not draw it!

(b) How many whites would a snail with 10 reds have?

(c) Copy and complete the rule for the snails.
number of white hexagons = ...

(d) Write the rule in shorthand. Use *w* for the number of white hexagons and *r* for the number of reds.

(e) Use your rule to work out the number of whites in a snail with 15 reds.

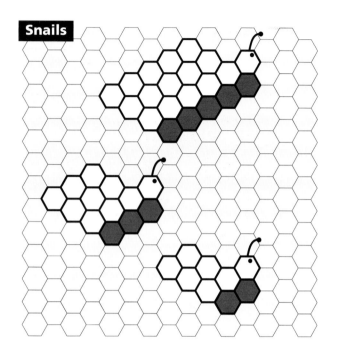

Snails

C2 These are slugs.

Slugs

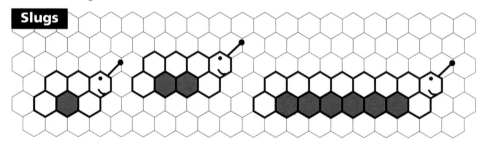

(a) How many white hexagons would you need for a slug with 10 reds? (Try to work it out without drawing.)

(b) How many whites would you need for a slug with 100 reds?

(c) One of the rules below works for the slugs. Which one is it?

$4w = r$ $w = 4r + 4$ $w = 4r$ $w + 4 = r$ $w = r + 4$

C3 These are larvae.

(a) Work out the missing numbers in this table.

r	2	3	4	5	6	7
w				13		

(b) How many whites would a larva with 10 reds have?

(c) How many whites would a larva with 1000 reds have?

(d) Write down a shorthand rule for the number of whites when you know the number of reds.

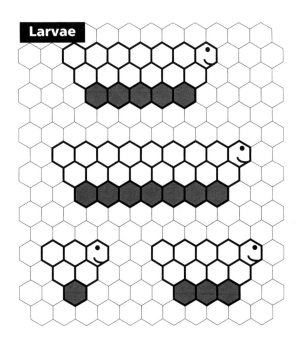

Larvae

C4 These are wigglers.

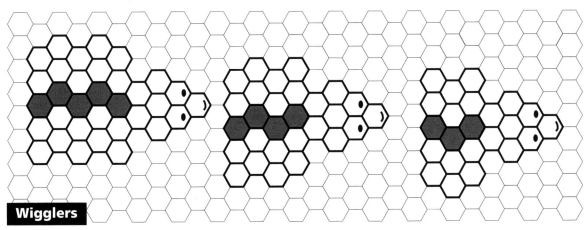

Wigglers

(a) How many whites are there in the wiggler with 5 reds?

(b) How many white hexagons would a wiggler with 6 reds have?

(c) How many whites would a 7 red wiggler have?

(d)

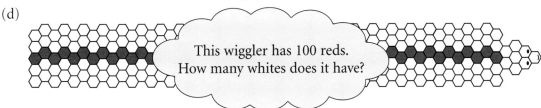

This wiggler has 100 reds. How many whites does it have?

(e) Write down a shorthand rule for the number of whites when you know the number of red hexagons in a wiggler.

C5 For each of these families of creatures,
work out which rule from the box is correct.

$w = 4r$ $w = r + 4$ $w = 2r + 3$

$w = 3r + 3$ $w = 3r + 2$ $w = 2r + 6$

Sea-snails

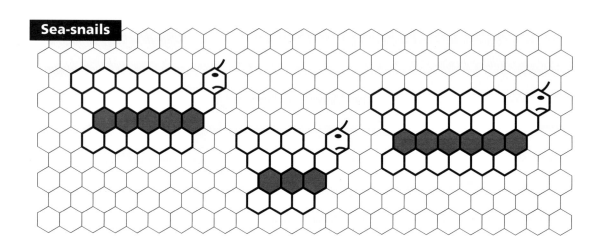

Sea-horses

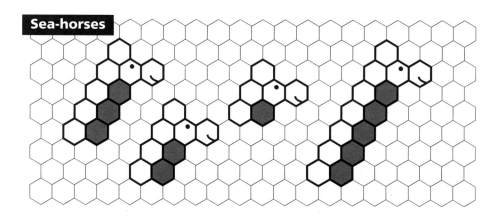

Flatfish

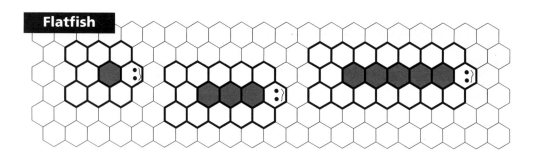

What progress have you made?

Statement

I can complete a table for a family of patterns.

Evidence

1 These are whelks.

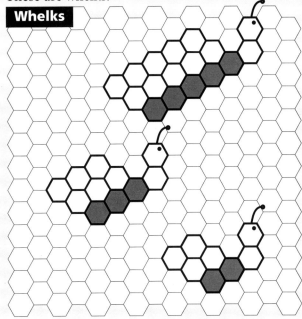

Whelks

(a) On sheet 226, draw a whelk with 4 red hexagons.

(b) Draw a whelk with 6 red hexagons.

(c) Without drawing, how many whites would there be in a whelk with 10 reds?

(d) Copy and complete this table.

Number of red hexagons	2	3	4	5	6	10	100
Number of white hexagons							

I can choose a rule for a family of patterns.

2 Which of these rules is correct for the whelk family?

number of white hexagons = number of red hexagons + 2

number of white hexagons = (number of red hexagons × 2) + 2

number of white hexagons = number of red hexagons × 2

I can write a rule using shorthand.

3 Write the rule for the whelk family in shorthand.

Use w for the number of white hexagons and r for the number of reds.

55

Surfaces of solids

This work will help you

♦ solve problems involving the nets of cubes and cuboids

♦ find the surface areas of cuboids

A Cubes

A flat shape which can be folded up to make a solid is called a **net** of the solid.

On squared paper, draw two different nets for a cube.

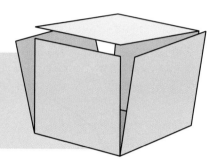

A1 A **hexomino** is a shape made by putting 6 squares together edge to edge. Which of the hexominoes on sheet 227 are nets of a cube?

A2 (a) Colour each square of hexomino A as shown.

(b) Cut out the net, and fold it into a cube.

(c) Check that whenever two faces meet at an edge they are different colours.

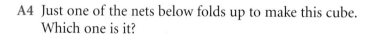

A3 Look at the other hexominoes on sheet 227 which are nets of a cube. Using 3 colours only, colour the faces of these nets so that all the cubes have different coloured faces meeting at every edge.

Now cut out the nets and check you were right.

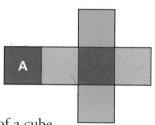

A4 Just one of the nets below folds up to make this cube. Which one is it?

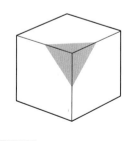

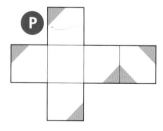 P

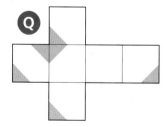 Q

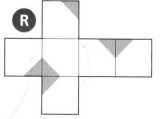 R

If you need to, check your answer by shading net A on sheet 228 and making the cube.

A5 This cube has a black face on the bottom and two flags on each of the four side faces. All the flags point to the left.

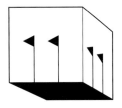

Draw the flags on nets B, C and D on sheet 228 so that each of them folds up to make the cube.

A6 This cube has three different bands going round it.

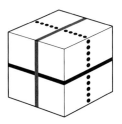

On nets E, F and G on sheet 228 draw lines so that each net can be folded to look exactly like this cube.

*A7 When you make solids from nets it is helpful to have **flaps** on some edges. Each flap is then stuck to another face. You have to make sure that you do not put two flaps on the same edge.

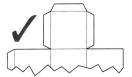

Which of these nets would stick together properly?

(a)

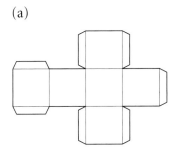

(b)

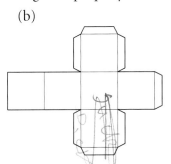

(c)

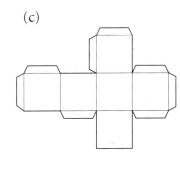

(d)

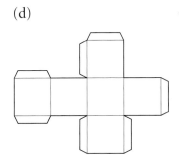

(e)

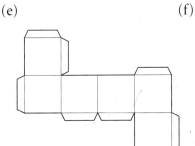

(f)

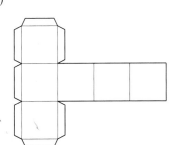

B Cuboids

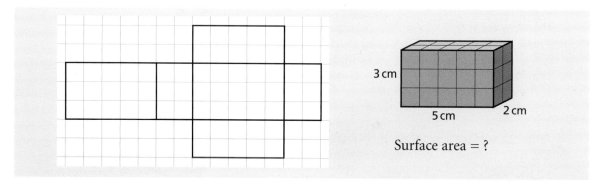

3 cm
5 cm
2 cm

Surface area = ?

B1 Here are four nets ...

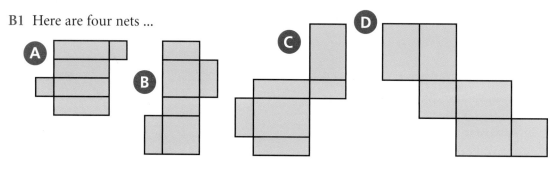

A

B

C

D

... and four cuboids

P

Q

R

S

Which net goes with
which cuboid?

B2 These are incomplete nets for
some cuboids.
One face is missing
from each net.

Draw each net on squared
paper and add the
missing face.

Compare your answers
with other people's.

If you need to, check your
answers by cutting out each
net and making the cuboid.

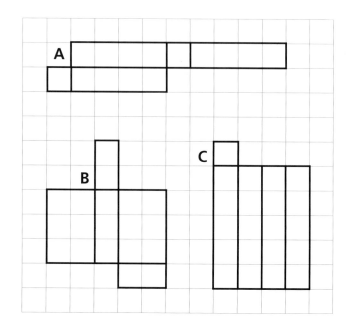

A

B

C

B3 (a) (i) What is the area of the front of this cuboid in cm² ?
(ii) What is the area of the back?

(b) Work out the area of the right side.

(c) Work out the area of the top.

(d) What is the total surface area of all six faces of the cuboid?

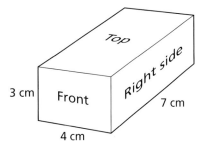

B4 Find the surface area of each cuboid.

(a)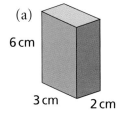

(b)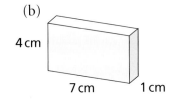

(c)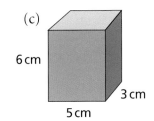

(d)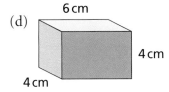

(e)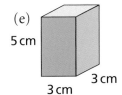

(f)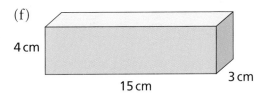

(g)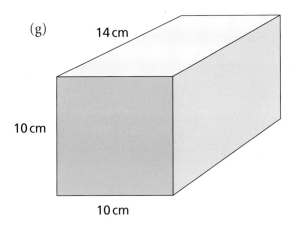

(h)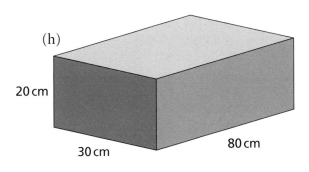

What progress have you made?

Statement

Evidence

I know when a hexomino is the net of a cube.

1 Which of these is **not** the net of a cube?

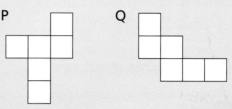

I can recognise nets of cuboids.

2 Here is a cuboid.

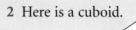

Which of these is a net of the cuboid?

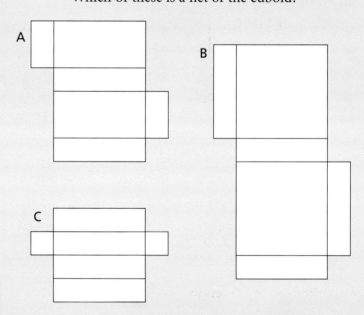

I can find the surface areas of cuboids.

3 Work out the surface area of this cuboid.

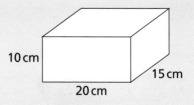

10 cm

20 cm

15 cm

⑪ Lost for words

These puzzles will help you practise finding fractions of whole numbers.

① Add the first $\frac{1}{4}$ of **LONESOME** to the last $\frac{1}{2}$ of **SAVE**.
- The first $\frac{1}{4}$ of the letters in **LONESOME** gives **LO**.
- The last $\frac{1}{2}$ of the letters in **SAVE** gives **VE**.

This makes **LOVE**.

② Add the first $\frac{1}{3}$ of **PRIVILEGE** to the first $\frac{1}{2}$ of **ZEAL**.
- The first $\frac{1}{3}$ of the letters in **PRIVILEGE** gives **PRI**.
- The first $\frac{1}{2}$ of the letters in **ZEAL** gives **ZE**.

This makes **PRIZE**.

Solve these puzzles

1   Add the first $\frac{1}{2}$ of SWEATY to the last $\frac{1}{2}$ of FEET.
2   Add the first $\frac{1}{3}$ of CLARET to the last $\frac{1}{2}$ of BLUE.
3   Add the first $\frac{1}{4}$ of IRON to the first $\frac{1}{3}$ of CEMENT.
4   Add the first $\frac{1}{5}$ of PEASHOOTER to the last $\frac{1}{2}$ of MENACE.
5   Add the first $\frac{1}{4}$ of EXERCISE to the first $\frac{1}{3}$ of AMAZON.
6   Add the first $\frac{1}{2}$ of SCHEME to the last $\frac{3}{4}$ of FOOL.
7   Add the last $\frac{1}{3}$ of ISOSCELES to the first $\frac{3}{5}$ of SONIC.
8   Add the first $\frac{2}{3}$ of CUT to the first $\frac{2}{5}$ of BEECH.
9   Add the first $\frac{2}{3}$ of HOLIER to the last $\frac{3}{5}$ of TODAY.
10   Add the first $\frac{3}{8}$ of DIAGONAL to the first $\frac{1}{2}$ of METEOR to the last $\frac{1}{3}$ of NUMBER.

Find the missing fraction

1   Add the first **?** of HELPED to the first $\frac{1}{3}$ of LONELY to give HELLO.
2   Add the first $\frac{3}{5}$ of PUPPY to the last **?** of TRANQUIL to give PUPIL.
3   Add the first **?** of PIZZICATO to the first $\frac{1}{4}$ of ZANZIBAR to give PIZZA.
4   Add the first $\frac{1}{3}$ of TOP to the last **?** of PREACHER to give TEACHER.

Do it yourself

Make up some fraction word puzzles of your own.

⑫ Buying and selling

This work will help you solve problems involving decimals.

A Perfect pizzas

COUNTY MARKET

How much is 1.4 kg of cheese?

What about 5.7 kg of olives?

Cheese £4.10 per kg
Olives £6.50 per kg
Anchovies £19.20 per kg
Mushrooms £3.40 per kg
Salami £13.90 per kg
Tomatoes £1.40 per kg

A1 Rosa's Café sells pizzas.
Here is Rosa's shopping list this week.

(a) How much will the cheese cost?

(b) Work out the cost of each item on the list.

(c) What will be the total cost?

> Cheese - 2.5 kg
> Salami - 1.8 kg
> Olives - 2.3 kg
> Anchovies - 0.3 kg
> Tomatoes - 9 kg
> Mushrooms - 4.6 kg

A2 Find the total cost for each of these lists.

(a)
> 1.8 kg of cheese
> 2.3 kg of salami
> 1 kg of olives
> 1.5 kg of anchovies
> 2.4 kg of mushrooms
> 7 kg of tomatoes

(b)
> Cheese - 1.4 kg
> Salami - 3 kg
> Olives - 0.8 kg
> Anchovies - 0.5 kg
> Mushrooms - 0.25 kg
> Tomatoes - 0.5 kg

A3 (a) Without using a calculator, which of these do you think would cost more?

2 kg of salami   5 kg of mushrooms

(b) Use your calculator to check.

B Rounding to the nearest penny

COUNTY MARKET

The cost of 3.3 kg of capers
is £6.79 × 3.3 = £22.407

How much is this?

Prawns £12.49 per kg

Capers £6.79 per kg

Carrots £0.33 per kg

Garlic £2.99 per kg

Pepperoni £14.05 per kg

Onions £0.52 per kg

2.4 kg onions
1.4 kg carrots
3.8 kg capers
2.9 kg onions
1.3 kg pepperoni
2.4 kg capers

B1 Round these amounts to the nearest penny.

(a) £6.713   (b) £2.678   (c) £4.239   (d) £10.342

(e) £4.725   (f) £1.408   (g) £7.135   (h) £1.203

(i) £2.419   (j) £3.297   (k) £1.195   (l) £2.997

B2 To the nearest penny, calculate the cost of each of these.

(a) 1.2 kg of capers   (b) 9.2 kg of onions   (c) 0.4 kg of pepperoni

(d) 5.5 kg of carrots   (e) 3.5 kg of capers   (f) 1.7 kg of carrots

(g) 1.4 kg of capers   (h) 2.7 kg of onions   (i) 2.1 kg of pepperoni

(j) 0.9 kg of capers   (k) 1.3 kg of onions   (l) 1.5 kg of carrots

B3 Copy each of these shopping lists.
Work out the cost of each item to the nearest penny.
Then work out the total bill for each list.

(a)

0.3 kg of capers
1.5 kg of pepperoni
6.3 kg of carrots
5.4 kg of onions
0.9 kg of prawns
0.3 kg of garlic

(b)

0.1 kg of capers
2.1 kg of onions
0.7 kg of garlic
3.4 kg of carrots
2.5 kg of pepperoni
2.1 kg of prawns

C Further rounding

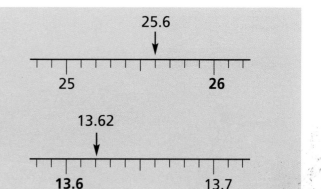

25.6 is between **25** and **26**.
It is nearer to **26** than 25.

When we round **25.6** to the
nearest whole number we get **26**.

When we round **13.62** to
one decimal place we get **13.6**.

C1 Write down, to two decimal places, what number each arrow below points to.

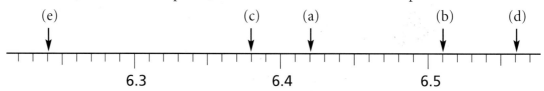

C2 Round each of your answers to question C1 to one decimal place.

C3 Round each of these to the nearest whole number.

    (a) 14.7   (b) 17.9   (c) 2.1   (d) 7.5   (e) 9.8

    (f) 0.8   (g) 99.7   (h) 49.1   (i) 103.5   (j) 109.9

C4 Round each of these to one decimal place (1 d.p.).

    (a) 3.67   (b) 4.11   (c) 110.66   (d) 0.17   (e) 9.55

    (f) 0.15   (g) 9.89   (h) 6.09   (i) 10.05   (j) 12.88

C5 (a) 5.99 is between 5.9 and 6.0 Which is it closer to?

    Round these to one decimal place.

    (b) 2.98   (c) 2.91   (d) 104.99   (e) 8.97   (f) 9.97

C6 SuperGold hi-fi cable costs £15.65 for 1 metre.
Work out the cost (to the nearest pound) of

    (a) 2 metres   (b) 4 metres   (c) 6 metres   (d) 12 metres

*C7 1 cubic centimetre (cm^3) of gold weighs 19.28 grams.
To one decimal place, work out the weight of

    (a) $2\,cm^3$ (b) $3\,cm^3$ (c) $21\,cm^3$ (d) $3.5\,cm^3$

What progress have you made?

Statement

I can solve problems involving decimals.

I can round to the nearest penny.

I can round to the nearest whole number and to one decimal place.

Evidence

1 Cheese costs £5.90 per kg.
   How much does 6.5 kg cost?

2 What is the cost of 3.3 metres of cloth at £7.90 per metre?

3 How much is 2.4 kg of ham at £3.50 per kg?

4 Round the following to the nearest penny.
   (a) £3.921   (b) £6.738   (c) £5.907
   (d) £4.203   (e) £1.965   (f) £8.597

5 (a) How much does 3.4 kg of meat cost at £3.79 per kg?
   (b) What is the cost of 1.8 metres of ribbon at £1.74 per metre?
   (c) How much is 2.5 kg of onions at £0.49 per kg?

6 Round these to the nearest whole number.
   (a) 45.2   (b) 5.7   (c) 10.6
   (d) 0.2   (e) 25.9   (f) 39.8

7 What number, to two decimal places, is each of these arrows pointing to?

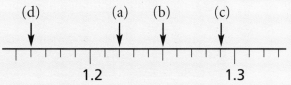

8 Write each of your answers to question 7 to one decimal place.

9 Round each of these to one decimal place.
   (a) 5.87   (b) 6.05   (c) 8.99
   (d) 0.26   (e) 0.09   (f) 19.96

⑬ Percentages

This work will help you

- ◆ understand what 'per cent' and 'percentage' mean
- ◆ calculate 50% and 25% of numbers and amounts of money
- ◆ change scores out of 5,10 or 20 into percentages

A How many per cent?

If you split a bar into 100 equal bits, each bit is 1%.

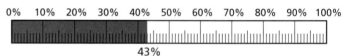

43%

A1   What percentage of each bar is red? What percentage is white?

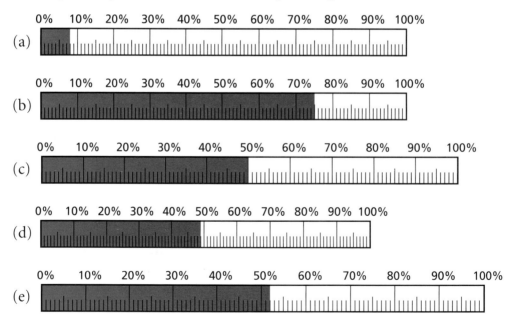

(a)

(b)

(c)

(d)

(e)

66

If you split a circle into 100 equal bits, each bit is 1%.

In this diagram 43% of the circle is shaded.

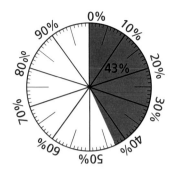

A2 In each diagram, what percentage of the circle is shaded? What percentage is not shaded?

(a)

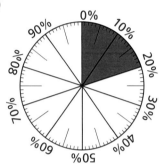

(b)

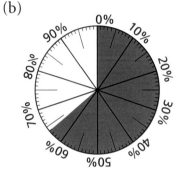

B Easy percentages

50 per cent of an amount is a half of it. To find a half of something, just divide by 2.

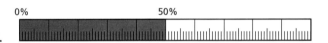

B1 Work these out.

(a) 50 per cent of 10p

(b) 50 per cent of £1

(c) 50 per cent of £60

(d) 50 per cent of £500

(e) 50 per cent of 74p

(f) 50 per cent of £2.40

Concert tickets

To reserve a seat pay 50% now and the rest on the day

B2 Work out what you have to pay now for

(a) one £10 ticket

(b) two £15 tickets

(c) three £6 tickets

(d) one £7 ticket

(e) one £3.60 ticket

(f) three £1.80 tickets

25 per cent of an amount is a quarter of it.
To find a quarter of something, just divide by 4.

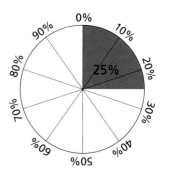

B3 Work these out.

    (a) 25% of 20p   (b) 25% of £1

    (c) 25% of £80   (d) 25% of £500

    (e) 25% of £100   (f) 25% of £1.20

CHEESE PROMOTION

Whatever you ask for you get 25% extra free

B4 How much do you get free when you ask for

    (a) 400 g of Cheddar cheese

    (b) 100 g of Parmesan cheese

    (c) 600 g of red Leicester cheese

    (d) 240 g of Stilton cheese

    (e) 2 kg of Edam cheese

    (f) 0.8 kg of Emmental cheese

B5 In each of these, find which is the greater amount.
Say how much it is.

    (a) 50% of £2 or 25% of £3.60   (b) 25% of £8 or 50% of £4.20

    (c) 50% of £1.20 or 25% of £2.80   (d) 25% of £12 or 50% of £5.98

    (e) 25% of 96p or 50% of 50p   (f) 50% of £90 or 25% of £200

C Shove the coin

Anjit is playing 'Shove the coin'.

From 10 cm away I had 100% success.

Then I tried from 20 cm away. I got 8 out of 10 in the circle.

From 30 cm away I got 5 out of 10 in the circle.

Scores out of 10 are easy to change into percentages.

If you split a circle into ten equal pieces ...

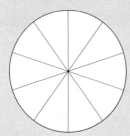

... each piece is 10%.

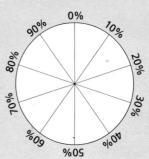

So, for example, 3 out of 10 is 30%.

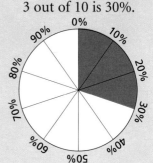

C1 Copy this table and finish it.

This means 4 out of 10 in the circle.

Score out of 10	0	1	2	3	4	5	6	7	8	9	10
Percentage	0%	10%				50%					100%

C2 Now play 'Shove the coin'.
Fill in a table like this for your scores.

Distance from the circle	10 cm	20 cm	30 cm	40 cm
Percentage in the circle				

C3 You need sheet 230.
   (a) Complete the statement above each circle.
   (b) Shade each circle to show the percentage.

C4 (a) Rose shoved the coin 10 times and got it in 3 times.
       What percentage was this?
   (b) Gill shoved the coin 10 times but got it in 7 times.
       What percentage was this?

C5 You need sheet 231.
   (a) Complete the statement above each circle.
   (b) Shade each circle to show the percentage.

C6 Change each of these into a percentage.
   (a) 4 out of 10   (b) 6 out of 10   (c) 0 out of 10
   (d) 4 out of 5   (e) 2 out of 5   (f) 5 out of 5

What progress have you made?

Statement

I know what 'per cent' and 'percentage' means.

Evidence

1 What percentage of this bar is shaded?

0%   50%   100%

I can calculate 50% and 25% of numbers and amounts of money.

2 Work these out.
   (a) 50% of 50p   (b) 25% of £200
   (c) 25% of 120   (d) 50% of 3 kg

I can change scores out of 5 or 10 into percentages.

3 Change these scores into percentages.
   (a) 7 out of 10   (b) 3 out of 10
   (c) 3 out of 5   (d) 1 out of 5

⑭ Extending still further

A Air traffic control

A game for two players

On squared paper, draw a playing grid like the one below.

Rules

The first person tells their partner where they want to go.	The partner then tries to place the cross. If they are wrong they lose the game.	The partner now tells the first person where they want to go.

The shortest route between a new point and any point on the grid must be **at least 3 units** along grid lines.

For example, if these points were already on the grid ...   ... this new point is OK ...   ... this one is not.

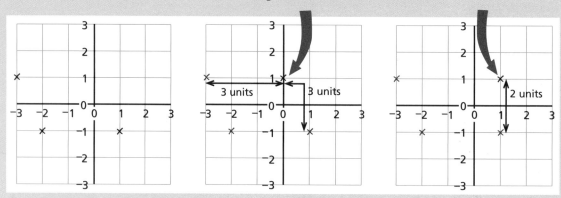

Now take it in turns to choose a point for your partner to place.
The winner is the last person to correctly choose where to put a cross on the grid.

B Going below

Here are some crosses
on a slightly larger grid.

B1 Write down the coordinates of
the crosses on the grid.

B2 The shortest route between
D and G along grid lines is 4.

How long is the shortest route
along grid lines between

(a) F and A

(b) G and B

(c) E and B

(d) $(1, 0)$ and $(3, 0)$ (e) $(^-1, ^-1)$ and $(3, ^-1)$ (f) $(^-3, ^-2)$ and $(^-1, ^-4)$

B3 For each of the quadrilaterals below only three pairs of coordinates are given.

Draw a grid going from $^-6$ to 6 in both directions.
Plot the coordinates given and complete each shape.
Write down the coordinates of the fourth point of each shape.

(a) Square $(3, ^-2)$ $(4, ^-4)$ $(6, ^-3)$

(b) Rectangle $(0, ^-2)$ $(2, ^-2)$ $(2, ^-5)$

(c) Parallelogram $(^-4, ^-2)$ $(^-1, ^-2)$ $(^-2, ^-4)$

(d) Kite $(^-1, 4)$ $(1, 3)$ $(^-1, ^-1)$

(e) Arrowhead $(6, 3)$ $(4, 1)$ $(4, ^-1)$

B4 On squared paper, draw an *x*-axis from $^-6$ to 6 and a *y*-axis from $^-7$ to 10.
Then plot the points below and join them up as you go along.

Start a new set of points for each box.

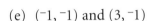

$(0, ^-2)$ $(1, ^-3)$ $(1, ^-4)$ $(0, ^-5)$ $(2, ^-5)$ $(1, ^-1)$

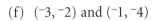

$(0, 9)$ $(0, 8)$

$(^-4, 8)$ $(^-4, 5)$ $(^-3, 3)$ $(^-1, 1)$ $(^-1, 4)$

$(5, 7)$ $(6, 7)$ $(6, 8)$ $(5, 8)$

$(^-5, ^-6)$ $(^-4, ^-5)$ $(^-3, ^-5)$ $(^-3, ^-3)$ $(^-2, ^-1)$ $(^-1, 0)$ $(0, 3)$ $(^-1, 4)$
$(^-2, 6)$ $(^-4, 8)$ $(^-3, 10)$ $(^-1, 10)$ $(0, 9)$ $(3, 9)$ $(5, 8)$ $(5, 7)$
$(4, 6)$ $(2, 6)$ $(1, 5)$ $(3, 0)$ $(3, ^-5)$ $(4, ^-5)$ $(5, ^-6)$ $(^-5, ^-6)$

⑮ Numbers go to town

This work will help you

◆ spot number patterns

◆ describe or explain number patterns

◆ continue number patterns

A Next door neighbours

B Short houses, tall houses

B1 In each of the streets below

• Which number is on the 5th tall house?

• Explain how you got your answer.

(a)

(b)

(c)

(d)

72

B2 In each of the streets below
- Which number is on the **10th** tall house?
- Explain how you got your answer.

(a)

(b)

(c)

(d)

B3 In each of the streets below
- Which number is on the 8th tall house?
- Explain how you got your answer.

(a)

(b)

(c)

(d)

(e)

(f)

B4 (a) Is 31 on a tall house in this street?
(b) How do you know ?

*B5** (a) Is 48 on a tall house in this street?
(b) How do you know ?

B6 In this street the numbers on the tall houses
form the sequence 3, 7, 11, ...

Copy these sequences and fill in the gaps.

(a) 3, 7, 11, __, __   (b) 2, 4, 6, 8, 10, __, __   (c) 2, 5, 8, 11, __, __

(d) 2, 7, 12, 17, __, __   (e) 1, 4, 7, 10, 13, __, __   (f) 3, __, 9, 12, 15, __

(g) __, __, 12, 16, 20   (h) __, __, 10, 14, 18, 22   (i) 1, __, __, 13, 17, 21

(j) 3, __, 15, __, 27   (k) 1, 2, 4, 7, 11, __, __   (l) 1, 2, 4, 5, 7, 8, __, __

(m) 2, __, 7, 10, 12, 15, 17, __   (n) __, __, 5, 8, 12, 17, 23

73

C Where do the numbers live?

2-row grid

1	3	5	7	9	11	13	row 1
2	4	6	8	10	12		row 2

3-row grid

1	4	7	10	13	16	19	row 1
2	5	8	11	14	17		row 2
3	6	9	12	15	18		row 3

4-row grid

1	5	9	13	17	21	25	row 1
2	6	10	14	18	22		row 2
3	7	11	15	19	23		row 3
4	8	12	16	20	24		row 4

5-row grid

1	6	11	16	21	26	31	row 1
2	7	12	17	22	27		row 2
3	8	13	18	23	28		row 3
4	9	14	19	24	29		row 4
5	10	15	20	25	30		row 5

C1 For each grid above
  (a) Which row is 37 in?
  (b) Explain how you know.

C2 In the 5-row grid, 7 and 17 are on either side of 12.

7	12	17

(a) Which numbers are either side of 59?

?	59	?

(b) Which numbers are either side of 82?

?	82	?

(c) Which numbers are either side of 101?

?	101	?

(d) Describe the pattern that helped you to answer (a), (b), and (c).

C3 (a) Copy this 5-row grid on to
        centimetre squared paper.
        Add two more columns to the grid.

     (b) How can you tell if a number is in
         a red or a yellow square?

1	6	11	16	21	26
2	7	12	17	22	27
3	8	13	18	23	28
4	9	14	19	24	29
5	10	15	20	25	30

C4 This is a 5-row grid.
    18 is in the 3rd row of the grid.
    Which row are these numbers in?

    (a) 47   (b) 68
    (c) 150   (d) 229

C5 Give three numbers between 70 and 90
    on the 4th row of this 5-row grid.

*C6 This is a 6-row grid.

    (a) What patterns can you see in each row?

    (b) How can you tell if a number is in a
        red or a yellow row?

    (c) In which row would you find these
        numbers?

        (i) 54   (ii) 45

    (d) Write three numbers between 50 and
        80 which are next to each other in a
        red row.

    (e) Write three numbers between 50 and
        80 which are next to each other in a
        yellow row.

1	7	13	19	25	31
2	8	14	20	26	32
3	9	15	21	27	
4	10	16	22	28	
5	11	17	23	29	
6	12	18	24	30	

D Squares

You need centimetre squared paper.

This diagram shows a 7 by 7 grid.

5 'big' squares have been coloured in.

This leaves12 grid squares not coloured.

9 + 16 + 4 + 4 + 4 = 37 coloured
plus 12 not coloured = 49 in total

Can you colour in different 'big' squares
so that fewer grid squares are left
not coloured?

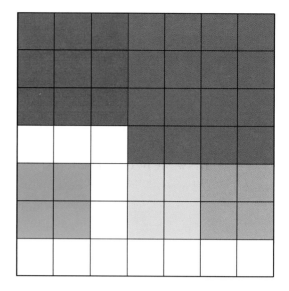

D1 Write the number of small squares in each large square.
These numbers are called **square numbers.**

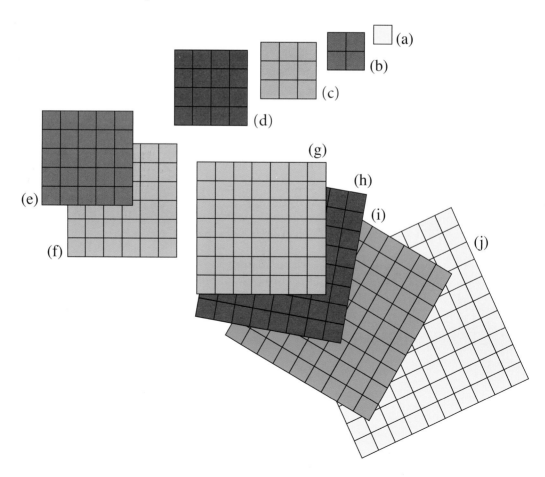

D2 Which of these numbers are square numbers?

64 88 36 100 144 50 121

D3 (a) Copy these triangle patterns.

(b) Write under each pattern the number of dots in it.
These numbers are called **triangle numbers**.

(c) Draw the next two patterns.
Write the number of dots in each pattern underneath.

(d) Without drawing, work out how many dots
there are in each of the next two patterns.
Draw them to check.

(e) The 10th pattern of triangle numbers has 55 dots.
How many dots will the 11th pattern have?

D4 What sort of numbers are on the diagonal
of this multiplication square?

×	1	2	3	4
1	1	2	3	4
2	2	4	6	⌣
3	3	6	9	1͟
4	4	8	12	16

D5 (a) Copy and complete these sums.

$$1 \qquad\qquad =$$
$$1 + 2 + 1 \qquad\qquad =$$
$$1 + 2 + 3 + 2 + 1 \qquad =$$
$$1 + 2 + 3 + 4 + 3 + 2 + 1 \quad =$$

(b) Add two more lines which continue the pattern.

(c) What pattern do the answers show?

(d) What will be the sum of the line where 8 is the middle number?

D6 For each of these patterns below
• copy and complete the sums shown.
• add two more lines to the pattern.
• describe the pattern in the answers.

(a) $1 \qquad\qquad =$
$1 + 3 \qquad\quad =$
$1 + 3 + 5 \qquad =$
$1 + 3 + 5 + 7 =$

(b) $(1 \times 2) \div 2 \ =$
$(2 \times 3) \div 2 \ =$
$(3 \times 4) \div 2 \ =$

(c) $(1 \times 2) - 1 =$
$(2 \times 3) - 2 =$
$(3 \times 4) - 3 =$

What progress have you made?

Statement	Evidence

I can continue number patterns.

1 Write the next two numbers in each of these.

   (a) 3, 6, 9, 12, __, __

   (b) 4, 7, 10, __, __

   (c) 6, 12, 18, 24, 30, __, __

   (d) 1, 5, 9, 13, 17, __, __

   (e) 103, 110, 117, 124, 131, __, __

   (f) 2, 5, 10, 17, 26, __, __

I can spot and describe number patterns.

2 Describe the number patterns in the rows, columns and diagonals of this grid.

```
 1
 4   6
 7   9  11
10  12  14  16
13  15  17  19  21
16  18  20  22  24  26
19  21  23  25  27  29  31
```

I can recognise the patterns of square numbers and triangle numbers.

3 Copy and complete these.
   Add two more rows to each pattern.
   Describe the patterns in the answers.

   (a) $(2 \times 2 - 2) \div 2 =$
   $(3 \times 3 - 3) \div 2 =$
   $(4 \times 4 - 4) \div 2 =$

   (b) $(1 \times 3) - (1 \times 2) =$
   $(2 \times 5) - (2 \times 3) =$
   $(3 \times 7) - (3 \times 4) =$
   $(4 \times 9) - (4 \times 5) =$

Review 2

1 These worms are made from red and white hexagons.

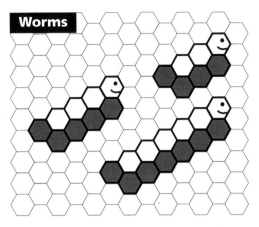

Worms

(a) Look at the worm with 5 red hexagons. How many whites does it have?

(b) Copy and complete this table for the worms. r stands for the number of red hexagons, w stands for the number of whites.

r	4	5	6	7	8
w					

(c) Which of these rules works for the worms? $w = r + 1$ $w = r - 1$ $w = 2r$

2 For each of these, say whether it is the net of a cube or not.

(a)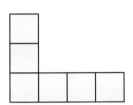

(b)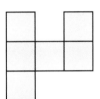

(c)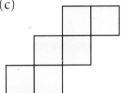

3 Find the surface area of each of these cuboids.

(a)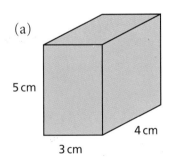

5 cm, 4 cm, 3 cm

(b)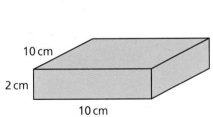

10 cm, 2 cm, 10 cm

(c)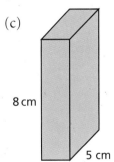

8 cm, 5 cm, 2 cm

4 What animal do you get in each of these fraction puzzles?

(a) Add the first $\frac{1}{2}$ of KITBAG to the first $\frac{1}{2}$ of TENNIS.

(b) Add the first $\frac{1}{3}$ of PURPLE to the last $\frac{1}{2}$ of SNAPPY.

(c) Take the last $\frac{1}{2}$ of WOODBINE from inside CABINET.

(d) Add the first $\frac{1}{4}$ of HORRIBLE to the last $\frac{1}{2}$ of COURSE.

(e) Add the first $\frac{3}{7}$ of DUCHESS to the first $\frac{5}{8}$ of KLINGONS.

5 Work out these costs.
   Give your answers to the nearest penny.

   (a) 1.4 metres of gold braid

   (b) 0.65 metres of plain cord

   (c) 1.75 metres of silk ribbon

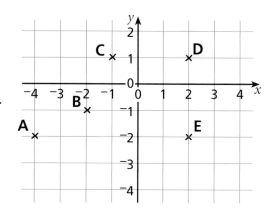

6 Work these out.

   (a) 50% of 120 grams   (b) 25% of 80 kg   (c) 50% of £4.80

7 (a) 3 of my 5 kittens are black. What percentage is this?

   (b) What percentage of my kittens are not black?

8 (a) Write down the coordinates of each
        of the points A, B, C, D and E.

   (b) Make a copy of the diagram and points
        on squared paper.

   (c) On your diagram, mark the point F ($^-$1, $^-$2).
        What do we call the shape
        with corners C, D, E and F?

   (d) On your diagram, mark G ($^-$3, $^-$4).
        What is the shape with corners
        C, D, E and G called?

9 In this street

   (a) What number is on the 10th tall house?

   (b) Explain how you got your answer.

10 In each of these, what number is on the 10th tall house? Explain how you know.

   (a)

   (b)

   (c)

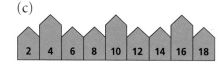

   (d)

11 Which of the numbers in the cloud are

   (a) odd numbers

   (b) square numbers

   (c) multiples of 4

   (d) factors of 45

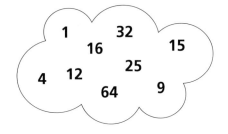

Mean

This work will help you find the mean for a set of data.

A Fair's fair

Ann's group

Name	Number of cans
Ann	2
Joe	6
Tina	3
Chinny	9

Ken's group

Name	Number of cans
Ken	9
Asif	2
Siobhan	7

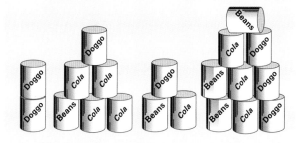

My group did better than yours because we collected more cans.

Ann

I think we did better.

Ken

Which group do you think did better?

A1 This table shows how many cans Sharon's group collected.

(a) How many cans did Sharon's group collect in total?

(b) Find the mean number of cans for Sharon's group.

Sharon's group

Name	Cans
Sharon	6
Amy	7
Rajit	8
Mark	7

A2 These tables show some results for other groups.

Find the mean number of cans for each group.

Amit's group

Name	Cans
Amit	10
Prakash	1
Jane	2
Debbie	1
Kay	1

Jason's group

Name	Cans
Jason	3
Isha	9
Purva	5
Nina	3
Rick	3
Helen	1

A3 Class 9J also collect cans for recycling.
These tables show how many they collected one week.

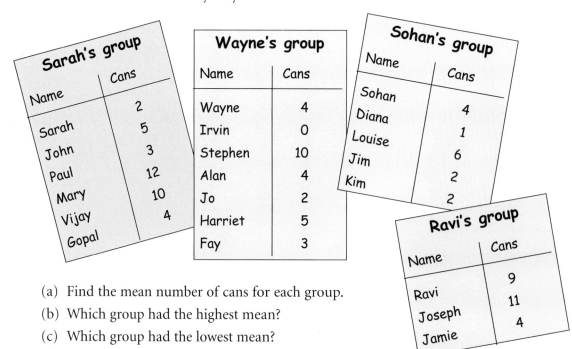

Sarah's group

Name	Cans
Sarah	2
John	5
Paul	3
Mary	12
Vijay	10
Gopal	4

Wayne's group

Name	Cans
Wayne	4
Irvin	0
Stephen	10
Alan	4
Jo	2
Harriet	5
Fay	3

Sohan's group

Name	Cans
Sohan	4
Diana	1
Louise	6
Jim	2
Kim	2

Ravi's group

Name	Cans
Ravi	9
Joseph	11
Jamie	4

(a) Find the mean number of cans for each group.

(b) Which group had the highest mean?

(c) Which group had the lowest mean?

A4 Class 9N collects bottles for recycling.
These tables show how many bottles they collected one week.

Group 1

	Bottles
Shane	13
Chris	10
Wing Yang	5
Sara	19
Martin	21
Lynn	6
Samantha	8
Jacob	14

Group 2

	Bottles
Adam	12
Patrick	13
Ryan	16
Jenny	9
Lee	0

Group 3

	Bottles
Andrew	10
Sam	12
Jamie	29
Lucy	1

Group 4

	Bottles
Leighanne	12
Michael	10
Delroy	11
Teresa	10
Clare	12

Group 5

	Bottles
Barry	16
Vicky	13
Katie	26
Sean	10
Jason	9
Sarah	22

Which group do you think each person belongs to?

(a) *My group had the highest mean.*

(b) *I didn't collect any bottles.*

(c) *Everyone in my group collected roughly the same number of bottles.*

(d) *We collected the same number of bottles as the group with the highest mean but our mean was lower.*

(e) *The mean for my group was 10 bottles.*

(f) *I collected the most bottles.*

83

A5 The table shows group 1's results for the week before.

Group 1	Bottles
Shane	5
Chris	2
Wing Yang	5
Sara	9
Martin	13
Lynn	7
Samantha	7
Jacob	12

The total number of bottles is 60.

So the mean number is 60 ÷ 8, which is 7.5.

But you can't have 7.5 bottles.

A mean of 7.5 is OK. It is just a number you can use to compare.

Here are the results for the other groups.

Group 2	Bottles
Adam	10
Patrick	2
Ryan	21
Lee	15
Jenny	9

Group 3	Bottles
Andrew	7
Sam	10
Jamie	8
Lucy	9

Group 4	Bottles
Leigh	14
Michael	8
Delroy	13
Teresa	19
Clare	7

Group 5	Bottles
Barry	9
Vicky	8
Katie	6
Sean	13
Jason	0
Sarah	4

(a) Find the mean number of bottles for each group.

(b) Which group had the highest mean?

(c) Which group had the lowest mean?

Pack it in

Do the makers always put exactly the same number of sweets in a packet?

- Count the number of sweets in some packets.

- You see 'average contents' on some packets. What does it mean?

- What would be a good number to use for 'average contents' on the packets you used?

B Comparing means

This activity is described in the teachers guide.

B1 Some classes do a memory experiment with ten pictures.

(a) This shows how many pictures each pupil in class 7Y remembered.

4 8 7 7 5 7 8 7 7 10 7 7 8 6 7 6 7 7 8 9

What is the mean number of pictures remembered by 7Y?

(b) These are the results for class 10T.

8 7 7 9 9 6 8 7 9 7 8 9 8 7 4 7 8 8 9 7 10 7 8 9 9

What is the mean number of pictures remembered by 10T?

(c) Which class do you think did better at remembering these pictures?

B2 The classes do a memory experiment with ten words.
This shows how many words each pupil in the two classes remembered.

7Y 7 5 6 7 5 6 9 7 7 5 6 3 6 7 7 8 4 7 6 8

10T 8 8 7 7 3 7 9 10 7 9 8 4 7 8 9 8 6 6 7 4 7 7 9 9 6

(a) Find the mean number of words remembered for each class.

(b) Which class did better at remembering the words?

*B3 Were the classes better at remembering pictures or words?
Give a reason for your answer.

C Decimal means

C1 This table shows the weights of some new-born baby girls.

(a) Who weighed 3.61 kg at birth?

(b) List the babies that weighed less than 3.5 kg at birth.

(c) What is the mean weight of these baby girls?

Girls	
Name	Weight (kg)
Rosanna	3.14
Ruth	3.29
Poppy	2.84
Ellen	3.61
Jenny	3.02
Faye	3.66

C2 This table shows the weights of some new-born baby boys.

(a) What is the mean weight of these baby boys?

(b) The mean birth weight of a baby boy in the UK is about 3.4 kg. Which of these baby boys weighed more than this?

Boys	
Name	Weight (kg)
Daniel	3.81
Andrew	4.22
David	2.96
Ben	3.26
Peter	3.85

C3 These tables shows the birth weights of some babies that were born two weeks earlier than expected.

Find the mean weight of

(a) the baby girls

(b) the baby boys

Girls	
Name	Weight (kg)
Kim	1.84
Estelle	2.76
Keisha	2.92
Jane	2.90
Clara	3.22
Phuong	4.27
Sue	3.42
Mira	2.41

Boys	
Name	Weight (kg)
Ravi	2.68
Adam	2.90
Rory	3.15
Chris	3.16
Stephen	2.38
Simon	2.98
Philip	4.13
David	2.94

C4 What can you say about the mean weights of the boys compared to the girls in question C3?

There is a National Seal Sanctuary in Cornwall.

It looks after rescued seals until they are well enough to be released back to the sea.

This table shows information about some young seals the sanctuary rescued.

Name	Reason for rescue	Weight at rescue (kg)	Weight at release (kg)
Lion	abandoned	15.4	50
Peace	malnourished	14.4	56
Yorkie	malnourished	16.4	100
Gurnard	trauma by nets	25.0	81
Shanny	malnourished	14.0	55
Montague	malnourished	13.1	84
Bass	malnourished	21.8	77
Mako	malnourished	15.6	65
Allis	malnourished	14.8	56
Mackerel	eye infection	21.8	66

Use the table on page 86 to answer these questions.

C5   What did Mako weigh when she was released?

C6   Which seals weighed less than 15 kg at rescue?

C7   Which seal was the lightest at rescue?

C8   How much weight did Bass gain between rescue and release?

C9   Which seals gained over 50 kg between rescue and release?

C10 What was the mean weight for the seals at rescue?

C11 What was the mean weight at release?

*C12 Some seals are not fit enough
     to survive on their own.
     They stay at the seal sanctuary.

     The tables show the approximate
     weights of these seals.

     Calculate the mean weight for

     (a) the male seals

     (b) the female seals

Male seals

Name	Weight (kg)
Benny	210
Spitfire	220
Charlie	225
Flipper	210
Scooby	200
Magnus	250

Female seals

Name	Weight (kg)
Fatima	160
Honey	150
Anneka	150
Silky	175
Sheba	165
Jenny	180
Twiggy	145

Aiming high a game for two players

You need	• A set of game cards (sheet 234)
The game	• Shuffle the cards and deal two cards to each player.
	• Place the rest of the cards face down.
	• Take turns to pick a card if you want to.
	• Each player can stop at any time.
	• Once one player has stopped, the other player can continue until he or she wants to stop or has 6 cards.
	• The maximum number of cards for one player is 6.
	• Once both players have stopped picking up cards the game is over.
The winner	• The player holding the set of cards with the highest mean at the end of the game

What progress have you made?

I can calculate the mean when the data are whole numbers.

1

Jen's group	
Name	Cans
Jen	7
Fiona	8
Halim	9
Don	4

Jim's group	
Name	Cans
Jim	2
Dawn	8
Kay	9
Jeff	8
Pete	3

What is the mean number of cans for

(a) Jen's group   (b) Jim's group

2 The table below shows the number of matches in some boxes.

Box	A	B	C	D	E
Number of matches	92	93	89	91	87

Find the mean number of matches for these boxes.

I can calculate the mean from data with decimals.

3 These tables show the weights of some one-year-old children.

Boys	
Name	Weight (kg)
Joe	10.8
Jake	13.6
Noah	9.8
Leroy	9.7
Tony	10.3
Anil	11.1
Halim	8.4
Kit	9.5

Girls	
Name	Weight (kg
Razia	9.8
Linda	10.7
Shirley	11.3
Mel	9.5
Victoria	10.1
Tara	8.7
Mira	9.9

Find the mean weight of

(a) the boys   (b) the girls

17 To scale

This is about understanding scales. The work will help you

♦ use a scale on a map or diagram to measure distances

♦ work out dimensions when you know the scale of a drawing

A Cutting down to size

This activity is described in
the teacher's guide.

This is a picture of Robert Wadlow.
He became the tallest man ever recorded.

This picture shows him in scout
uniform when he was 11 years old.
He was about 208 cm tall!

This picture makes him 8 cm tall.

How many times larger was he in real
life than in this picture?

On sheet 235 draw yourself and some
friends to the same scale.

How many times larger ...

... are these objects in real life?

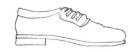

B Using scales

Maps and drawings often have a scale marked on them.

To find a measurement between two points:

- Mark the points on the edge of some scrap paper.
- Place the edge against the scale.
- If it is more than 50 m, mark the paper and slide along.

Boeing 777

B1 Use the scale to find these measurements.

    (a) The width of the tail (marked T on the plan)

    (b) The width of the fuselage (marked F on the plan)

    (c) The wingspan (the distance between the wing tips)

    (d) The length of the plane down the middle

B2 This is a map of Great Britain.

Use the scale to answer these questions.

(a) How far is it from London to

    (i) Leicester

    (ii) Liverpool

    (iii) Edinburgh

(b) A small aircraft has enough fuel to fly 500 km.
Can it fly from Liverpool to Dublin and back?

(c) Many people have run, cycled walked etc. from Land's End to John O'Groats.

How far is this 'as the crow flies' (in a straight line)?

Scale in kilometres

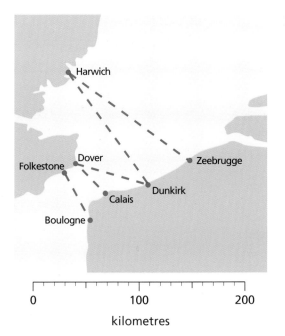

kilometres

B3 This map shows some Channel ferry routes.
The scale is shown below the map.
Use the scale to answer these questions.

(a) How many kilometres is the ferry route from Folkestone to Boulogne?

(b) How far is the route from Dover to Calais?

(c) (i) Is it further from Harwich to Dunkirk, or from Harwich to Zeebrugge?

    (ii) How much further?

Use sheet 236 to answer questions B4 to B6.

B4 Romeo's Pizzas is in East Cowes on the Isle of Wight.
The delivery charge is based on how far away you live.
All distances are measured in a straight line.

What would the delivery charge be
for someone living in

(a) Newport   (b) Sandown

(c) Yarmouth   (d) Ventnor

Romeo's Pizzas
Delivery Charges
Up to 5 miles - Free!
From 5 to 10 miles - £5
Over 10 miles - £10

B5 A canoe race is held around the Isle of Wight.
The race starts at Shanklin.

Work out the shortest distance round the island and back to Shanklin.

B6 Janet is organising a cycle ride on the Isle of Wight.
She wants the ride to start and end in Newport and to be about 30 miles.

Draw a route on your map that she could use.

You need sheets 237 and 238 for the next question.

B7 Rita is designing a new kitchen.
Sheet 237 shows a plan of her kitchen when empty.
Sheet 238 shows things she can buy for her kitchen from her DIY shop.

However, there are a few rules.

- The kitchen door must be able to open to 90°.
- The sink unit must be somewhere over the plumbing.
- The cooker must be somewhere over a socket.
- The washing machine must be no more than 2 m away from the plumbing.
  and no more than 1.5 m from a socket. The same applies to the dishwasher.
- All the cupboards have doors which must have room to open.

(a) Cut out one of the single cupboards on sheet 238.
How long would the cupboard be in real life?

(b) How long would one of the double cupboards be in real life?

(c) How long would the sink unit be in real life?

(d) Cut out all the other things on sheet 238.
Use them to plan a design for Rita's kitchen.
(You don't need to use everything on the sheet!)

Stick your design down on the plan.

Redesign your kitchen at home

Measure your own kitchen and make a plan using the scale on sheet 237.
Fit some of the units from sheet 238 into your own kitchen.

C One to one

How far between the Snorings?

You need sheet 239

- How far is it between Great Snoring and Little Snoring?
- There is a steam railway running from Little Walsingham to Wells-next-the-sea. How long is it?
- What village is 6 km west of Wighton?

Use the map on sheet 239 to answer these questions.

C1 How far is it in a straight line, to the nearest kilometre, from
    (a) North Creake to Wighton   (b) Stanhoe to Bircham Tofts
    (c) Fring to Burnham Market   (d) East Barsham to Great Snoring

C2 There is an island north of Brancaster Staithe.
How long is this island from east to west?

C3 Name the places which are
    (a) 5 km south-west by road from Stanhoe
    (b) 2 km north of South Creake by road
    (c) 6 km due east of Great Bircham
    (d) 8 km north-east of Docking

C4 A cyclist is planning a tour in this area, starting at Little Snoring.
He plans his route as follows.

Leg A   From Little Snoring go north through Great Walsingham and Warham All Saints to the A149.

Leg B   Go along the A149 to Brancaster.

Leg C   Go south down the B1153 to Docking.

Leg D   Go southeast on the B1454 to the A148.

Leg E   Along the A148 back to Little Snoring.

Measure each leg of his tour and find the total distance cycled.

C5 This is a map of Chris's garden.

It uses a scale where 1 cm stands for 1 m.

(a) How far is it from the tree to the nearest corner of the patio?

(b) How long and wide is the shed?

(c) Chris's dad wants to make a path from the patio to the shed.
How long will it need to be?

(d) They have a hose 5 m long. Will it reach from the tap to

(i) the pond

(ii) the furthest end of the flower bed

(iii) the tree

(e) What are the length and width of the patio?

(f) What is the area of the patio?

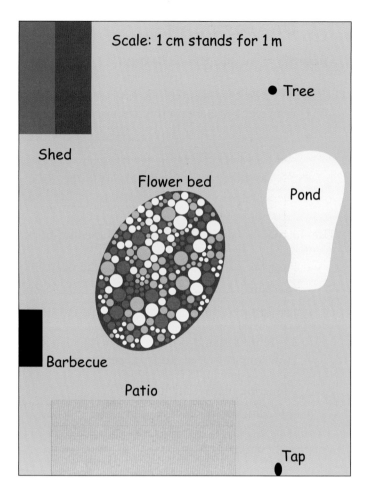

Scale: 1 cm stands for 1 m

Tree

Shed

Flower bed

Pond

Barbecue

Patio

Tap

You need sheet 240 for the next question.

C6 Midfolk County Council want to build a new roundabout.
A plan of the new design is shown on sheet 240.
It uses a scale where 1 cm stands for 1 m.

They want to make sure that vehicles can go round the roundabout without hitting the kerb or the roundabout.

(a) Here are the dimensions of some vehicles.
Cut out some scale models of these vehicles.
Which of these can go round the roundabout?

Car	4.2 m by 1.6 m
Ambulance	5.5 m by 2.6 m
Bus	10.4 m by 3.0 m
Fire engine	8.5 m by 3.0 m

(b) Design your own road and roundabout system on a piece of plain paper.
It should be possible for all the vehicles above to go round it.
How wide is your roundabout?

D Different scales

The right pitch?

Did you know that football pitches vary quite a lot in size?
This is a diagram showing the size of Manchester United's pitch at Old Trafford.

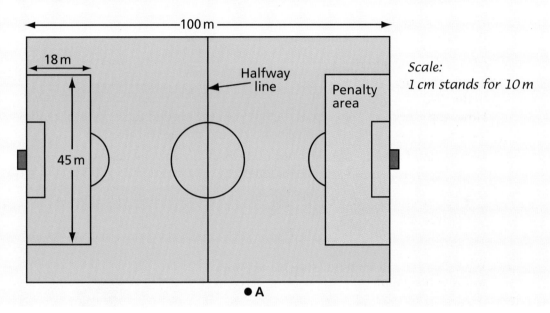

Answer these questions from the diagram above.

D1 (a) How many metres is it from the halfway line to the edge of the penalty area?

(b) How wide is the goal area (the rectangle in front of the goal)?

(c) How far is it from a corner to the edge of the nearest goal?

D2 A player at A can throw the ball 30 metres.

Can he reach the penalty area on (a) his right (b) his left

D3 (a) In training, footballers often run around the pitch. How far is this?

(b) If a fit person can run 100 m in 13 seconds,
roughly how long will one circuit take?

How big are your pitches?

You will need a long tape measure for this activity.

Measure the pitches (for any game) at your school or local recreation area.
Are they all the same size?

Make a scale drawing of one of the pitches you have measured.
Use a scale where 1 cm stands for 10 m.

D4 This shows the side view of a Concorde aircraft.
It is drawn to a scale where 1 cm stands for 4 m.

(a) What do these measurements stand for on the view?

    (i) 2 cm   (ii) 5 cm   (iii) 10 cm   (iv) 2.5 cm

(b) Measure the height of the top of the tail from the ground to the nearest centimetre.
How high is the top of the tail from the ground on the real aircraft?

(c) Measure the distance from the front to the back door to the nearest centimetre.
What is this distance on a real Concorde?

(d) Measure the length of the aircraft on the view to the nearest half-centimetre.
How long is the real Concorde?

D5 This shows the Empire State Building in New York.
It is drawn to a scale where 1 cm stands for 50 m.

(a) Measure the height of the building in the
view to the nearest centimetre.

(b) Use your answer to part (a) to find the
real height of the Empire State Building.

How high would these real objects
be if they were shown on the same scale?

(c) 'Big Ben' in London, nearly 100 m tall

(d) Blackpool Tower, 150 m tall

(e) The Eiffel Tower, 325 m tall

(f) The CN Tower in Toronto, 550 m tall

(g) Ben Nevis, the highest mountain in
the British Isles – just under 1350 m

(h) Mount Everest – 8 850 m

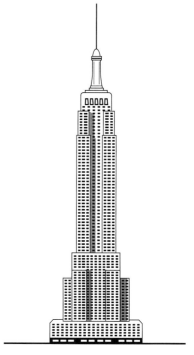

Design a poster

There are several Internet sites which tell you the heights of the world's tallest buildings.
Look these up and use them to design a poster called 'The World's Tallest Buildings'.

What progress have you made?

Statement

I can find measurements using scales.

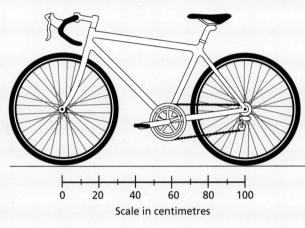

Scale in centimetres

Evidence

1 Use the diagram of the bike to find

    (a) the diameter of the wheels (including the tyres)

    (b) the height of the saddle above the ground

    (c) the distance between the centres of the wheels

I can read information from a scale drawing.

2 Here is a scale drawing of a computer. It uses a scale where 1 cm stands for 5 cm.

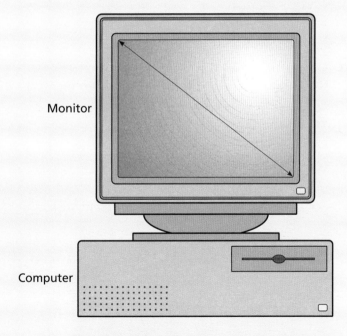

    (a) What is the height of the real computer and monitor?

    (b) What is the width of the real computer?

    (c) What is the width of the real disc slot?

    (d) How long is the diagonal of the screen?

⑱ Questions and calculations

This is about knowing which calculation to do to answer a problem.

A Add, subtract, multiply or divide?

Calculations

4 – 15	15 ÷ 12	15 × 4
15 + 12	4 × 12	15 – 12
15 + 4	12 + 4	15 × 12

15 pens

12 + 3	15 ÷ 12	15 – 12
15 ÷ 3	12 – 3	12 × 3
12 ÷ 3	12 + 15	15 – 3

12 pencils

16 – 2	2 ÷ 16	10 × 4
16 + 10	16 – 10	10 + 4
16 ÷ 2	16 + 2	16 × 10

16 poster paints

6 + 24	24 ÷ 6	24 – 4
24 × 4	4 ÷ 24	24 – 6
6 – 24	24 + 4	24 ÷ 4

24 paint sticks

13 – 4	4 ÷ 13	13 + 3
13 × 3	13 – 3	13 ÷ 3
13 ÷ 4	13 × 4	13 + 4

13 water colours

Write down which calculation opposite goes with each question below.
Then find the answer to the calculation.

Questions

A1 Pat takes 12 pens out of the art case.
    How many are left?

A2 Ken has 4 pens. He buys an art case.
    How many pens has he got altogether now?

A3 How many pens would there be altogether in 4 art cases?

A4 Sue, Raj and Kim share equally the pencils in one case.
    How many do they each have?

A5 There are 3 pencils missing from Zak's case.
    How many pencils are in Zak's case?

A6 Chris has 15 pencils. How many will not fit in the case?

A7 Jim has a full art case. He buys 10 more poster paints.
    How many poster paints has he now?

A8 A teacher has 10 full cases. How many poster paints has he got?

A9 In a full case there are two rows of poster paints.
    How many paints are there in each row?

A10 There are 6 blue paint sticks in a full case.
     How many are not blue?

A11 Pupils at a table share equally the paint sticks from a full case.
     They get 6 paint sticks each. How many pupils are at the table?

A12 A quarter of the paint sticks are missing from Amy's case. How many are missing?

A13 Nita has a full art case and another case with only 4 water colours left in it.
     How many water colours has she got?

A14 If you bought three art cases, how many water colours would you get?

A15 Jake finds that three of the watercolours in his set are used up.
     How many watercolours does Jake have left?

B Your own questions

Make up five questions about the picture.

B1 Look carefully at the picture below.

Make up five questions about the picture.
Write the calculation for each of your questions.

Now work out the answer to each of your questions.

⑲ Spot the mistake 1

In each piece of work there is one mistake.

Spot each mistake and correct it.
Try to say why the mistake was made.

Add, subtract, multiply and divide

1 Write in figures
three thousand and forty-eight.

> 300048

2 Calculate 43 + 39.

$$\begin{array}{r} 4\ 3 \\ +\ 3\ 9 \\ \hline 7\ 1\ 2 \end{array}$$

3 Calculate $10 - 2 \times 3$.

> $10 - 2 \times 3 = 24$

4 Find the missing number in $* - 9 = 3$.

> 9 - 6 = 3 so the missing number is 6

5 Use your calculator to calculate $\frac{20 + 8}{4}$.

> $20 + 8 \div 4 = 22$

6 Calculate 93 − 47.

$$\begin{array}{r} 9\ 3 \\ -\ 4\ 7 \\ \hline 5\ 4 \end{array}$$

Handling data

1 This bar chart shows the favourite colours for class 9Y.

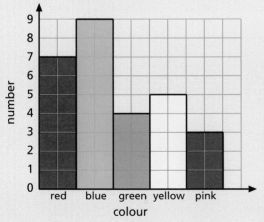

What is the mode for this chart?

> The tallest bar shows 9 people like blue best so the mode is 9.

2 The number of bottles collected by a group of pupils were

John 5   Paul 9   Spencer 3
David 0   Sue 3

What was the mean number of bottles collected?

> Mean = $\frac{5 + 9 + 3 + 3}{4}$ = 5

3 The pointer in this spinner is spun round.

What is the probability of the arrow pointing to red?

> Probability of red is $\frac{1}{3}$

Fractions and decimals

1 What fraction of the rectangle is shaded?

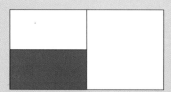

> The rectangle is split into 3 pieces.
>
> One piece is shaded so $\frac{1}{3}$ is shaded.

2 From the list of numbers

   0.7, 0.5, 0.12, 0.9, 0.3, 0.5

   what is the largest number?

> 0.12

3 Calculate 12.6 + 3.

> 12.6
> + 3
> ――――
> 12.9

4 Claire's class went on a trip in minibuses.
   There are 25 children in her class.
   Each minibus holds 10 children.

   How many minibuses were needed?

> 25 ÷ 10 = 2.5
>
> so 2.5 minibuses were needed

5 Put these numbers in order, smallest first.

   0.03, 0.2, 0.5, 0.04, 0.06

> 0.2, 0.03, 0.04, 0.5, 0.06

Area

1 What is the area of this rectangle?

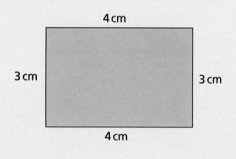

> Area = 4 + 3 + 4 + 3 = 14 cm

2 What is the area of this shape?

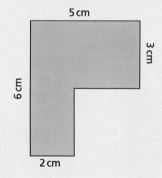

> Area = 5 × 3 × 2 × 6 = 180 cm²

Tens, hundreds and thousands review

This work will help you multiply and divide numbers by 10, 100 and 1000.

A Big numbers

Words and numbers

Match the numbers and words.
Say the headlines for each of these using the words below.

A 20 000 see United crash

B Girl wins £8200

C Pop star leaves £820 000

D 2 000 000 homes built

E £82 000 paid for a shed

F 200 000 pack the park

G 8 200 000 tax dodgers

H 2000 wedding guests

two hundred thousand

eighty-two thousand

two thousand

eight hundred and twenty thousand

two million

eight thousand two hundred

twenty thousand

eight million two hundred thousand

Big bingo

Choose five numbers from this list.
Write them down.
This is your bingo card.

60 000	300 000	4 000 000
3 000 000	50 000	600 000
400 000	5 000 000	30 000
6 000 000	40 000	500 000

Saying numbers

1 (a) 71 000   (b) 70 100   (c) 70 105   (d) 70 001

2 (a) 39 200   (b) 30 920   (c) 30 902   (d) 30 092

3 (a) 45 680   (b) 40 568   (c) 45 068   (d) 45 006

4 (a) 267 000   (b) 260 700   (c) 260 070   (d) 260 007

5 (a) 206 700   (b) 206 070   (c) 206 007   (d) 200 067

B Multiplying by 10, 100, 1000

Multiplying by 10 moves digits one place to the left.
A zero is put in the units place.

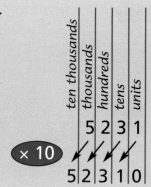

This diagram shows what happens when
you multiply 5231 by 10.

So $5231 \times 10 = 52\,310$

B1 Write down the results of these multiplications

(a) 35×10 (b) 3×10 (c) 125×10 (d) 10×350

B2 Match these multiplications with their answers.

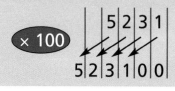

$45 \times 10 =$ $405 \times 10 =$

$4500 \times 10 =$ $450 \times 10 =$ $4 \times 10 =$

4500   4050   40   45 000   450

Multiplying by 100 moves digits
two places to the left.

Multiplying by 1000 moves digits three
places to the left.

B3 Match these multiplications to numbers in the box.
(You won't use all the numbers in the box.)

4000   1200   30 000
400 000   35 000
1 000 000   120   350 000

(a) 12×10 (b) 35×1000 (c) 100×12

(d) 1000×1000 (e) 10×120 (f) 40×100

B4 Work these out.

(a) 33×100 (b) 1000×24 (c) 103×10 (d) 100×40

(e) 306×1000 (f) 100×210 (g) 251×100 (h) 7002×1000

B5 Copy and complete these multiplications.

(a) $42 \times 10 = \square$ (b) $52 \times \square = 5200$ (c) $120 \times \square = 1200$

(d) $89 \times \square = 89\,000$ (e) $\square \times 100 = 74\,000$ (a) $\square \times 10 = 7000$

C Dividing by 10, 100, 1000

Dividing by 10 moves digits one place to the right.

Dividing by 100 moves digits two places to the right.

Dividing by 1000 moves digits three places to the right.

C1 Match these divisions to numbers in the box.

(a) 2500 ÷ 10   (b) 70 000 ÷ 100   (c) 2500 ÷ 100

(d) 80 000 ÷ 1000   (e) 700 ÷ 10   (f) 800 ÷ 100

> 25   8   80
> 700   250   70
> 7   800

C2 Work these out.

(a) 6000 ÷ 10   (b) 57 000 ÷ 100   (c) 2400 ÷ 100   (d) 67 000 ÷ 10

(e) 290 000 ÷ 1000   (f) 4000 ÷ 100   (g) 25 000 ÷ 1000   (h) 7 000 000 ÷ 100

C3 Work these out. They are a mixture of multiplications and divisions.

(a) 100 × 23   (b) 4600 ÷ 10   (c) 730 × 100   (d) 4800 × 100

(e) 6000 ÷ 100   (f) 10 200 ÷ 10   (g) 70 × 1000   (h) 24 000 ÷ 100

C4 Find the missing numbers.

(a) 4 × 100 = ...   (b) 5000 ÷ ... = 5   (c) 10 × ... = 30   (d) 700 ÷ 100 = ...

(e) ... × 100 = 1500   (f) ... × 10 = 450   (g) 14 × 1000 = ...   (h) 280 ÷ 10 = ...

(i) 20 × 100 = ...   (j) 4200 ÷ ... = 42   (k) 4000 ÷ 100 = ...   (l) ... × 100 = 3000

C5 Copy and complete these chains.

(a) (30) →×10→ ○ →÷100→ ○ →×1000→ ○

(b) (70) →÷10→ ○ →×1000→ ○ →÷100→ ○

(c) (43) →×100→ ○ →÷10→ ○ →×1000→ ○

(d) (7000) →÷10→ ○ →×100→ ○ →÷1000→ ○

C6 Work out each of these.
Then put them in order of size, smallest first.

> 3700 ÷ 10   52 × 10   8500 ÷ 10
> 100 000 ÷ 1000   23 × 100

*C7 Write these on seven pieces of paper.

Arrange them to make a correct chain.
Stick or copy your chain in your book.

> 480   480 000   4800   48
> ÷10   × 100   × 1000

D Multiply or divide?

D1 A box contains 12 tins of cat food.
How many tins will there be in 10 boxes?

D2 A block of flats has 10 floors.
There are 60 flats altogether in the block.
How many flats are there on each floor?

D3 A box of tissues contains 150 tissues.
How many are there altogether in 10 boxes?

D4 In a raffle 2000 people each bought 10 tickets.
How many tickets did they buy altogether?

D5 It will cost £450 000 to rebuild a bridge that fell down after a flood.
How many donations of £10 will be needed?

D6 3500 collectors for a charity have 100 stickers each to give out.
How many stickers do they have altogether?

D7 At a school fair, people made a line of 3000 pennies.
How many pounds was this?

D8 An advertising company wants to send out 150 000 letters.
They buy envelopes in boxes of one thousand.
How many boxes will they need?

What progress have you made?

Statement

I can read and write large numbers.

I can multiply and divide whole numbers by 10, 100, 1000.

Evidence

1 Write these in figures.
   (a) Forty thousand and fifty
   (b) Two hundred and three thousand and six

2 Work these out.
   (a) 350 × 100
   (b) 42 000 ÷ 100
   (c) 1000 × 460
   (d) 30 200 ÷ 10

21 Solids

This work will help you

◆ describe, make and draw three-dimensional objects made from cubes

◆ find the number of cubes in larger shapes

A Describing

For pupils working in pairs

Each person in the pair has some multilink cubes.

Sit back to back. No looking round!
One of you makes an object with the cubes.
Then describe it to your partner.
Your partner tries to make it.

When your partner has finished,
compare your objects.

B Drawing 3-D objects

You need triangular dotty paper.

Here is how you can draw a cube using dotty paper.
First make sure your dotty paper is the right way round.

Draw one face on a slope.

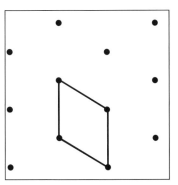

Draw lines going back.

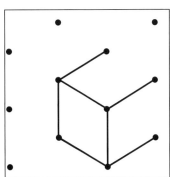

Finish the back face.

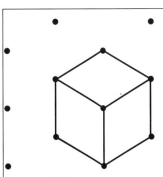

Practise drawing some cubes on dotty paper.
Make them different sizes.

Here is a multilink letter L.

This is a three-dimensional drawing of it.

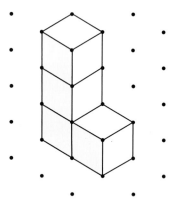

B1 Here are three more letters that have been started.

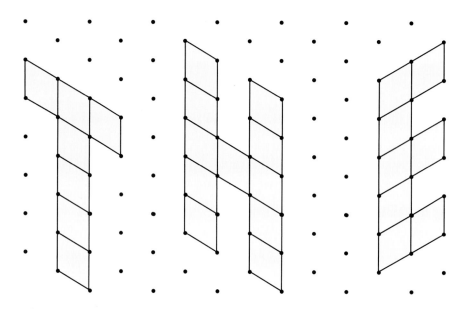

(a) Copy and complete these letters. Make them one cube thick.

(b) Draw a letter F in the same way.

(c) Draw three other letters.

B2 Use dotty paper to draw the shapes in these photos.

(a)
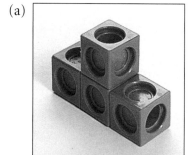

(b)
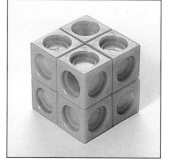

(c)
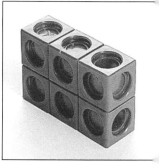

B3 This is a drawing of a multilink shape.

Afzal has started drawing the shape from a different direction.

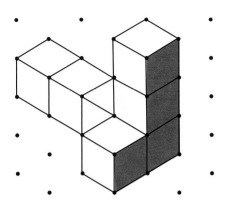

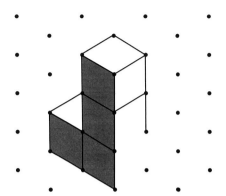

Copy and complete Afzal's view of the shape.

B4 This question is on sheet 241.

B5 Look at this multilink shape.
Some of the drawings below are drawings of this shape. Some are not.

Write down which are correct drawings.

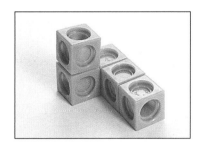

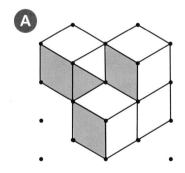

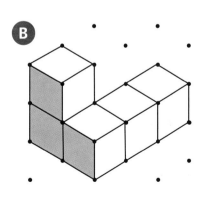

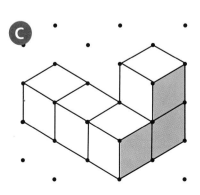

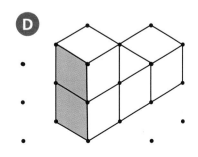

C Cuboids

Which of these objects are shaped like cuboids?

How would you tell someone what a cuboid was?

C1 This cuboid is made from 12 cubes.
It is a '6 by 2 by 1' cuboid.

Three other cuboids can be made
using all 12 cubes.

List all the other cuboids.

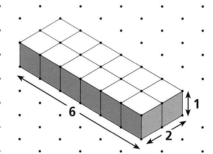

C2 List all the different cuboids that can
be made from 8 cubes.

C3 How many different cuboids can be made from

(a) 20 cubes   (b) 24 cubes

Keep a record of your work.

D Counting cubes

D1 Which of these cuboids has the most cubes? Which has the smallest?

Write them down in order of size, smallest first.

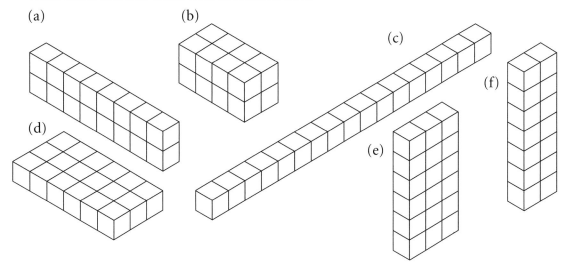

D2 How many cubes are there in each of these cuboids?

(a)   (b)   (c)   (d)

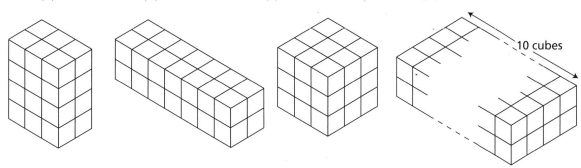

D3 Work out how many cubes there are in each of these.

(a)   (b)   (c)   (d)

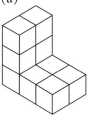 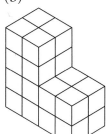 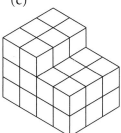 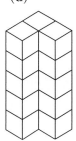

What progress have you made?

Statement

I can draw three-dimensional objects on triangular dotty paper.

Evidence

1 Draw this object on dotty paper.

I know what a cuboid is.

2 List three objects which are cuboids.

I can recognise cuboids.

3 Which of these shapes are cuboids?

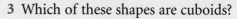

A
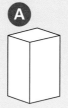

B
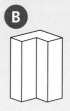

C

I can work out how many cubes there are in a cuboid.

4 How many cubes are there in each of these cuboids?

(a)
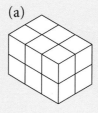

(b)
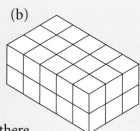

I can work out how many cubes there are in other shapes.

5 How many cubes are there in each of these shapes?

(a)
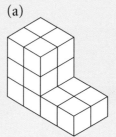

(b)
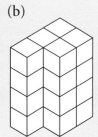

Review 3

1 These tables show the weights at birth of a litter of puppies.

(a) Work out the mean weight of all the puppies in the litter.

(b) Work out the mean weight of the female puppies.

(c) Find the mean weight of the male puppies.

(d) Which were heavier on average, the male or the female puppies?

Name	Male or female	Weight in grams
Dusty	F	163
Rusty	M	181
Minky	F	128
Merry	F	174
Becky	F	121
Spot	M	154
Jimmy	M	136

2 This map shows part of Europe. Use the scale below the map to find how far it is in a straight line from

(a) London to Paris

(b) London to Dublin

(c) London to Madrid

(d) Paris to Rome

3 (a) Maria flies from Rome to Madrid and back. How many kilometres does she travel?

(b) Hamish flies from Glasgow to Rome and back. How far does he travel?

4 Pick the calculation that goes with each problem.

(a) Anna buys 8 oranges that cost 24p each. How much do they cost altogether?

(b) Ben buys 8 chews for 24p. How much does each chew cost?

(c) Chloe has 24p and buys a pencil for 8p. How much money does she have left?

(d) Dave had some money and spent 8p. Now he has 24p left. How much did he have originally?

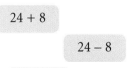

24 + 8

24 − 8

24 × 8

24 ÷ 8

5 Spot the mistake in each of these. Correct the mistake
 and say why you think it was made.

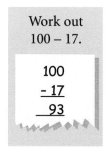

Work out
100 – 17.

100
– 17
93

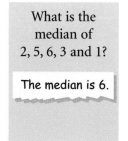

What is the
median of
2, 5, 6, 3 and 1?

The median is 6.

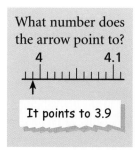

What number does
the arrow point to?
4   4.1

It points to 3.9

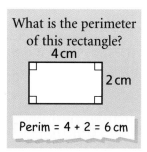

What is the perimeter
of this rectangle?
4 cm

2 cm

Perim = 4 + 2 = 6 cm

6 Rewrite each of these headlines using figures for the numbers.

 (a) Three hundred thousand fans greet team

 (b) **Two thousand and thirty breakdowns on railways**

 (c) *A million and one ways to win!*

 (d) Three hundred thousand two hundred and seven umbrellas lost

7 Without using a calculator work these out.

 (a) 22 × 100   (b) 43 000 ÷ 100   (c) 1000 × 41

 (d) 5000 ÷ 10   (e) 450 000 ÷ 100   (f) 201 × 100

8 Work these out without using a calculator.

 (a) 200 people each collect £100 for charity. How much do they collect altogether?

 (b) A group of 10 friends decide to buy a bus that costs £3000.
  How much will they each have to pay?

 (c) There are 100 raffle tickets in a book. Paige sells 20 books.
  How many tickets does she sell?

 (d) A carton of fresh duck eggs contains 10 eggs.
  I want to buy 50 eggs. How many cartons do I need?

9 How many cubes are there in each of these shapes?

 (a)   (b)   (c)   (d)

   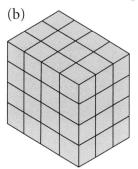   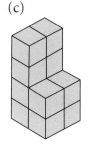   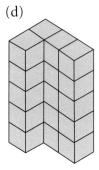

10 Which of the solids in question 9 are cuboids?

22 Decimals and metric units

This work will help you

◆ multiply and divide decimals by 10, 100, and 1000

◆ change between metric units

A Multiplying decimals by 10

How wide is this nut in millimetres?
How long is the row of nuts in millimetres?

The nut is 0.7 cm wide.

How long is the row of nuts in centimetres?

A1 This screw is 0.8 cm long.

10 of these screws are put in a row.
How long is this row of 10 screws?

A2 Write down the answers to these.
(a) 0.7 × 10 (b) 0.9 × 10 (c) 0.2 × 10 (d) 0.1 × 10

A3 This bolt is 1.3 cm long.

1.3 cm

How long is this row of 10 bolts?

A4 How long would a row of 10 bolts be if each bolt measured
(a) 1.5 cm (b) 1.8 cm (c) 2.5 cm (d) 3.0 cm (e) 12.5 cm

A5 Write down the answers to these.
(a) 1.7 × 10 (b) 1.9 × 10 (c) 5.2 × 10 (d) 10.5 × 10 (e) 25.4 × 10

B Multiplying and dividing decimals by 10 and 100

To multiply a decimal by 10 all the digits are moved one place to the left.

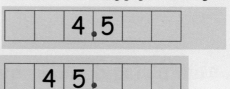

So 4.5 × 10 = 45

You can use tracing paper over a grid.

B1 Multiply each of these numbers by 10.

    (a) 5.9   (b) 23.4   (c) 0.3   (d) 8.6   (e) 10.4

B2 Multiply each of these numbers by 10.

    (a) 1.45   (b) 27.62   (c) 0.74   (d) 12.63   (e) 5.04

Dividing by 10 moves all the digits one place to the right.

 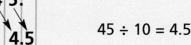

45 ÷ 10 = 4.5

B3 Divide each of these numbers by 10.

    (a) 65   (b) 406   (c) 8   (d) 50   (e) 702

    (f) 13.3   (g) 6.4   (h) 103.8   (i) 4.0   (j) 60.5

B4 Work these out.

    (a) 54 ÷ 10   (b) 4.2 × 10   (c) 14.6 × 10   (d) 14.6 ÷ 10

    (e) 10 × 3.07   (f) 13.82 ÷ 10   (g) 0.67 × 10   (h) 240 ÷ 10

Multiplying by 100 moves all the digits **two** places to the left.
Dividing by 100 moves all the digits **two** places to the right.

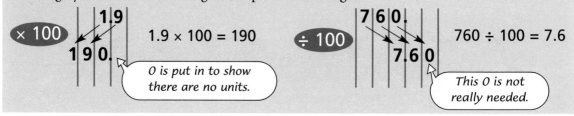

B5 Work these out.

    (a) 100 × 7.04   (b) 34.5 ÷ 100   (c) 0.97 × 100   (d) 51.8 ÷ 100

    (e) 0.6 × 100   (f) 100 × 1.08   (g) 32 ÷100   (h) 0.03 × 100

B6 Copy and complete these by writing a number or × or ÷.

    (a) $10 \times 6.3 = \square$ (b) $3.1 \times \square = 310$ (c) $420 \square 10 = 42$

    (d) $0.89 \times 10 = \square$ (e) $6.07 \times \square = 607$ (f) $0.03 \times \square = 3$

    (g) $4 \square \square = 0.04$ (h) $0.86 \square \square = 86$ (i) $\square \div 100 = 0.402$

B7 This code gives a number for each letter.

A	B	C	D	E	H	I	M	O	P	R	S	T	U	Z
7000	0.035	0.07	0.35	0.7	0.805	3.5	7	8.05	35	70	80.5	350	700	805

Work out the answers to these and use the code to change them to letters.
Rearrange each set of letters to spell a food.

    (a) $7 \div 10$ (b) 100×0.7 (c) 3.5×10 (d) $80.5 \div 10$

        $8.05 \div 10$ $3.5 \div 10$ 100×8.05 7×100

        8.05×10 0.07×10 $35 \div 10$ 3.5×10

        $0.7 \div 10$ 70×100 80.5×10 10×8.05

        $70 \div 100$ $3.5 \div 100$ $700\,000 \div 100$

        0.07×10

B8 Copy and complete these chains

    (a) 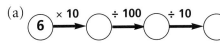   (b)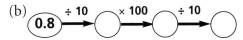

    (c) 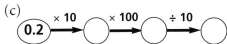   (d)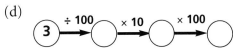

B9 Write these numbers in order of size, smallest first.

     650 ÷ 100   47 ÷ 10

   0.42 × 10   0.6 × 10   0.02 × 100

*B10 Write these on seven pieces of paper.

    Arrange them to make a correct chain.
    Stick or copy your chain in your book.

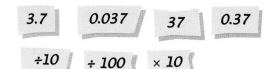

B11 Sarah has ten bottles of wine.
    Each bottle contains 0.7 litre.
    How many litres of wine does she have altogether?

B12 A machine is putting drugs into medicine bottles.
    7.2 g of one drug is shared equally into 100 bottles.
    What weight of drug goes into each bottle?

C Metric units

How long is this screw in centimetres?
How long is this screw in millimetres?

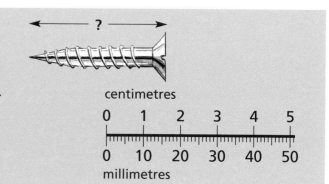

There are 10 millimetres in a centimetre.

To change centimetres to millimetres
you **multiply by 10**.
To change millimetres to centimetres
you **divide by 10**.

C1 Measure the lengths of these screws and write each length in

- centimetres
- millimetres

(a)    (b)

(c)    (d)    (e)

C2 Change these to millimetres.

(a) 7.4 cm   (b) 11.5 cm   (c) 7 cm   (d) 20 cm   (e) 0.6 cm

C3 Change these to centimetres.

(a) 40 mm   (b) 42 mm   (c) 138 mm   (d) 500 mm   (e) 9 mm

C4 What are the missing numbers in this table?

Animal	Length in cm	Length in mm
Pygmy shrew	8.6	(a)
Dormouse	(b)	139
Badger	82	(c)

C5 A teacher asked a group of pupils to measure the width of their index fingers.
Some of them measured in centimetres and some measured in millimetres.
Here are the results.

Sanjay 21 mm   Lizzie 19 mm   Martin 2 cm

Aaron 1.6 cm   Shari 1.8 cm

Write their names in order of size, smallest first.

There are 100 centimetres in a metre.

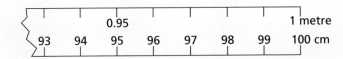

To change metres to centimetres you **multiply by 100**.
To change centimetres to metres you **divide by 100**.

C6 Change each of these to centimetres.

(a) 3 m   (b) 2.5 m   (c) 0.5 m   (d) 0.75 m   (e) 0.08 m

C7 Change each of these to metres.

(a) 600 cm   (b) 140 cm   (c) 35 cm   (d) 70 cm   (e) 9 cm

C8 (a) Change 2.05 m to cm.   (b) Change 107 cm to m.

     (c) Change 68 cm to m.   (d) Change 0.02 m to cm.

C9 What are the missing numbers in this table?

Animal	Length in centimetres	Length in metres
Grass snake	125	(a)
Stoat	38	(b)
Coypu	(c)	1.2
Weasel	(d)	0.25
Water vole	(e)	0.3
Pygmy shrew	9	(f)

How tall?

Zeng Jinlian, the world's tallest ever woman, was 248 cm tall when she died aged 17.
What is that in metres?

The world's shortest recorded woman was Pauline Musters who was only 61 cm tall when she
died aged 19. What is this height in metres?

How tall are you in   • centimetres
                   • metres

D Multiplying and dividing decimals by 1000

There are 1000 metres in a kilometre.

1 kilometre
1000 metres

To change kilometres to metres you **multiply by 1000**.
To change metres to kilometres you **divide by 1000**.

D1 Change each of these distances to metres.

    (a) 4 km   (b) 3.7 km   (c) 0.4 km   (d) 20 km   (e) 0.08 km

D2 Change each of these distances to kilometres.

    (a) 7000 m   (b) 6500 m   (c) 250 m   (d) 400 m   (e) 20 m

D3 Work these out.

    (a) 3.45×1000 (b) $270 \div 1000$ (c) 1000×31.3 (d) 0.06×1000

There are 1000 grams in a kilogram.

1 kilogram
1000 grams

Changing between grams and kilograms
is just like metres and kilometres.

D4 (a) Change 1.2 kilograms to grams.   (b) Change 1300 grams to kilograms.

    (c) Change 650 g to kg.   (d) Change 2.35 kg to g.

D5 Change the weights of these animals to kilograms.

    (a) Cat 3300 g   (b) Chinchilla 425 g

    (c) Hedgehog 785 g   (d) Mole 75 g

D6 Change the weights of these animals to grams.

    (a) Rabbit 2.5 kg   (b) Rat 0.28 kg

E Changing between units

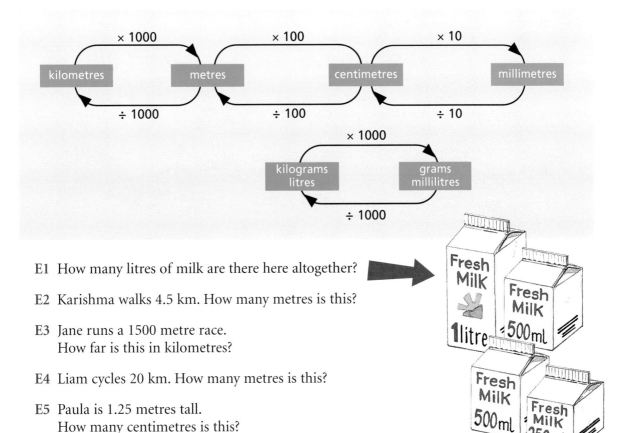

E1 How many litres of milk are there here altogether?

E2 Karishma walks 4.5 km. How many metres is this?

E3 Jane runs a 1500 metre race.
How far is this in kilometres?

E4 Liam cycles 20 km. How many metres is this?

E5 Paula is 1.25 metres tall.
How many centimetres is this?

E6 Rahela's baby is 65 cm long.
What is this length in metres?

E7 John's desk is 1.08 metres high.
How many centimetres is this?

E8 How many kilograms of cornflakes are
there here?

E9 (a) Change 5.3 kg to grams.

(b) Change 850 g to kilograms.

(c) Change 1400 millilitres to litres.

(d) Change 3.75 litres to millilitres.

Oral questions

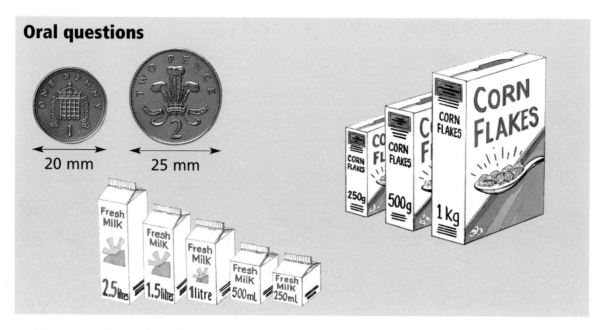

F Tens, hundreds and thousands review

Sheet 242 has questions which involve multiplying or dividing by 10, 100 or 1000.

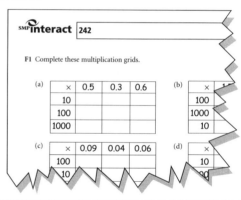

SMP interact 242

F1 Complete these multiplication grids.

(a)

×	0.5	0.3	0.6
10			
100			
1000			

(b)

×	1...
100	
1000	
10	

(c)

×	0.09	0.04	0.06
100			
10			

(d)

×	
10	

What progress have you made?

Statement	Evidence
I can multiply and divide decimals by 10, 100, 1000.	1 Work these out. (a) 7.5×10 (b) 100×3.45 (c) $65.3 \div 10$ (d) 3.27×1000 (e) $328 \div 100$ (f) $575 \div 1000$
I can change between metric units.	2 Change (a) 506 cm to m (b) 1.4 litres to ml (c) 5670 m to km (d) 3.25 kg to grams

㉓ Estimation and units

This work will help you

◆ estimate a wide range of distances

◆ convert between metric and imperial measures

A Good judgement

How long is one centimetre?

The nail of your index finger is about 1 cm across.
Check this with a ruler.

- Use your fingernail to find the length of your pencil.
  Check your estimate with a ruler.

- Use your fingernail to estimate the lengths of some other
  things on your desk.

How long is one metre?

- Hold a piece of chalk in each hand.
  Try to mark out 1 m using the chalk on a large piece of paper.
  Ask some others to do the same using different colour chalks.
  Now get a metre rule and see who is closest.

- Mark a line on the ground.
  Ask everyone to place a counter exactly 2 m away.
  Check who was closest with a tape measure.
  Try estimating 3 m and 5 m.

- Estimate, without measuring, the size of your room, the
  corridor, the size of your desk and other items around you.

How long is 100 metres?

- Mark a line, or use any available one, on a field or playground.
  Everyone walks to where they estimate 100 m away is.
  Mark the estimates and then measure to see who is closest.

- About how big are the sports pitches at your school?
  Estimate some other distances around the school.
  Use a trundle wheel or long tape measure to check how good your
  estimates were.

How far is a mile?

Make a list of places you think are exactly one mile from your school.
Check your guesses with a map.
Estimate how far you and your friends have to travel to school.
Make a list of local places and estimate how far they are away from your school.

Do not measure for these problems. Estimate each one.

A1 How far will you have to walk to your next lesson?
Estimate the distances for all of your walks between lessons today.

Roughly how far do you walk in a school day?

A2 If workbenches were put around the walls of your classroom,
how many chairs could you fit along the edge?

A3 How wide does a car parking space need to be?
Estimate the length of a suitable wall in your school.
How many car parking spaces could fit along this wall?

A4 The height of a horse is measured from the ground to the shoulders.
This is usually measured in 'hands'.
A 'hand' is roughly the width of a person's hand.

Measure the width of your hand in centimetres.
Roughly how high are these breeds of horse in centimetres?

(a) Arab (15 hands)   (b) Shire (17 hands)   (c) Suffolk Punch (16 hands)

A5 The Skylon Tower in Toronto has 154 floors.
Estimate the height of your classroom.
Use this to estimate the height of the Skylon Tower.

Handy measurements

In ancient times people based measurements
on parts of their body.
This picture shows some common ones.

They used these rules;
    1 cubit = 6 palms = 24 digits
    1 span = 3 palms
Check whether the rules work for your
own body measurements.

The Bible tells us that Noah's Ark was
300 cubits long.
Roughly how long was this in metres?

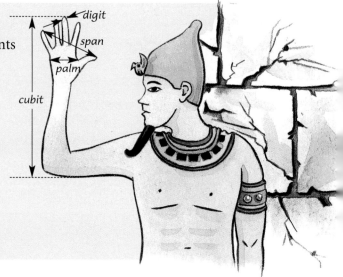

B Converting

One inch is roughly $2\frac{1}{2}$ cm. | 2 inches?

3 feet are about one metre. | 12 feet?

5 miles are about 8 kilometres. | 10 miles?

B1 1 inch (in) is about $2\frac{1}{2}$ centimetres.
About how long are each of these woodscrews in centimetres?
They are not drawn to scale!

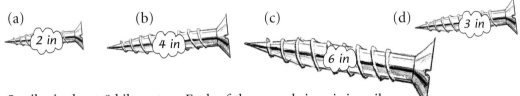

(a) 2 in

(b) 4 in

(c) 6 in

(d) 3 in

B2 5 miles is about 8 kilometres. Each of these road signs is in miles.
Convert each sign into kilometres.

Hyde	10 miles
Soan	20 miles

London	50 miles
Brighton	25 miles

B3 My kitchen measures 12 feet long by 6 feet wide by 9 feet high.

(a) What are my kitchen's dimensions in metres, roughly?

(b) Ceiling tiles cost £5 per square metre.
About how much will it cost to tile the ceiling?

One ounce is about 30 grams. | 3 ounces?

One pound is about $\frac{1}{2}$ a kilogram. | 4 pounds?

B4 One ounce (oz) in weight is about 30 grams.
Write out this old recipe using grams.

Scones
2 oz butter
4 oz sugar
8 oz flour
$\frac{1}{2}$ oz baking powder

4 lb   8 lb   10 lb   20 lb

B5 1 pound (lb) is about $\frac{1}{2}$ a kilogram.
About how many kilograms
do each of these weigh?

T

One pint is about
2 pints? $\frac{1}{2}$ a litre. 6 pints?

One gallon is
2 gallons? about $4\frac{1}{2}$ litres. 10 gallons?

B6 A group of friends are regular blood donors.
Their logbooks show how many pints of blood they have given so far this year.
Change these into litres.

(a) 4 pints (b) 8 pints (c) 10 pints (d) 5 pints

B7 How many litres does each of these barrels hold?

(a)  4 gallons
(b)  6 gallons
(c) 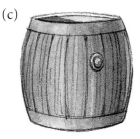 8 gallons

B8 Tim owns an old Mini. It holds 6 gallons of petrol.
Jill owns an old Citroen 2CV. It holds 30 litres of petrol.

Which car holds less petrol, the Mini or the 2CV? Explain.

B9 (a) Write out this recipe so that it is all in metric units.

(b) To make the soup, Mary buys a 5 kg sack of potatoes.

About what weight of potatoes are left after she makes the soup?

(c) Mary buys a 50 gram packet of parsley. About how much parsley does she have left?

Potato and leek soup
(serves six)

2 pounds white leeks
1 pound potatoes
2 pints chicken stock
½ pint cream
½ ounce chopped parsley
salt and pepper to taste

❖ *Trim, wash and chop the leeks into one inch pieces.*
❖ *Boil with the potatoes for 30 minutes in the chicken stock.*
❖ *Cool slightly and add cream and parsley.*
❖ *Can be served hot or cold.*

B10 (a) Will 2 pints of milk fit into a 2 litre jug?

(b) Will a 30 mm nail go right through a one inch plank?

(c) A lorry is 9 feet tall. Will it go under a bridge that is 3 m 50 cm high?

(d) Will a kilogram of tomatoes be enough for a recipe that needs 4 pounds?

(e) Will 5 gallons of beer fit into a 25 litre barrel?

Explain each of your answers.

C Graphs

You can use a graph to convert more exactly from one unit to another.
This graph can be used to convert between kilograms and pounds.

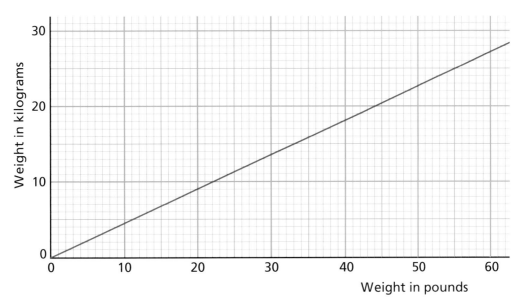

C1 (a) Use the graph to check that 22 pounds is the same as 10 kilograms.

(b) From the graph, check that 20 kilograms works out to be 44 pounds.

C2 Use the graph to convert each of these weights into kilograms.

(a) 9 pounds   (b) 33 pounds   (c) 51 pounds   (d) 11 pounds

C3 Use the graph to convert each of these weights into pounds.

(a) 9 kilograms   (b) 21 kilograms   (c) 25 kilograms   (d) 14 kilograms

C4 When she was born, Chloe weighed 4 kg.
When her mother was born, she weighed 10 pounds.
Which baby weighed more, Chloe or her mother?

C5 When Short Jack Silver buried his treasure there was 60 pounds of gold.
When it was found 100 years later, only 20 kg of gold was left.
How many kilograms of gold was missing?

C6 Which is the heavier, 56 pounds of potatoes or 20 kg?
What is the difference in weight, in kilograms?

C7 This question is on sheet 243.

What progress have you made?

Statement

I can estimate lengths.

Evidence

1 Which of these is the best estimate of the height of a normal door in a house?

   1 metre   2 metres   4 metres   8 metres

2 Which of these is the best estimate of the length of the foot of an adult?

   10cm   25cm   50cm   100cm

I can convert between metric and imperial units.

3 Without looking back in this book, check that you know (and write down)

   (a) how many grams there are in 1 ounce

   (b) how many feet there are in 1 metre

   (c) how many kilometres are in 5 miles

   (d) how many litres there are in 1 gallon

4 Change the following.

   (a) 3 ounces into grams

   (b) 15 miles into kilometres

   (c) 2 inches into centimetres

   (d) 10 pints into litres

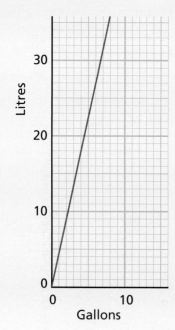

5 Use the graph on the left to answer these questions.

   (a) My motorbike holds 4 gallons of petrol. Can I put 15 litres into an empty tank?

   (b) Cath's fish tank needs 5 gallons of water to fill it. Will 20 litres be enough?

6 Read the graph as accurately as you can to convert

   (a) 30 litres to gallons

   (b) 2 gallons to litres

   (c) 6.5 gallons to litres

(24) Spot the mistake 2

In each piece of work there is one mistake.

Spot each mistake and correct it.

Try to say why the mistake was made.

Measures

1 The time is 14:45.

What time will it be in 20 minutes?

> 14:45 + 20 = 14:65

2 A bag of sugar weighed 1 kg, but 20 grams of sugar were lost through a hole in the bottom.

How much sugar is left in the bag?

> 100 – 20 = 80
>
> so 80 grams are left in the bag

3 The time is now 3:00.

What was the time 50 minutes ago?

> 3:00 – 50 = 2:50

4 Write 4 metres 7 centimetres in metres.

> 4.7 metres

5 Change 2000 metres to kilometres.

> 2000 ÷ 100 = 20

Coordinates

1 What are the coordinates of point C?

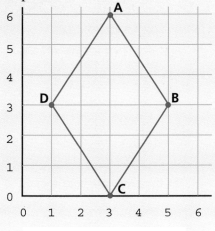

The coordinates are (0, 3).

2 On a grid, plot the points (3, 5), (4, 8) and (6, 10).

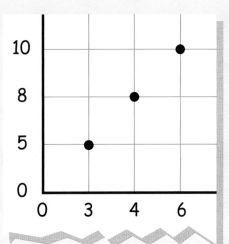

129

Angles

1 Measure this angle in degrees.

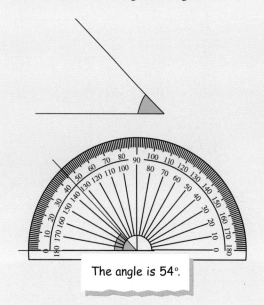

The angle is 54°.

2 Measure this angle in degrees.

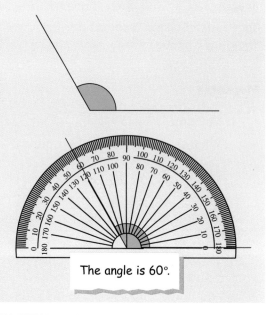

The angle is 60°.

Symmetry

1 Draw a rectangle.
Show clearly any lines of symmetry.

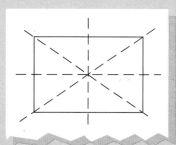

2 Draw a parallelogram.
Show clearly any lines of symmetry.

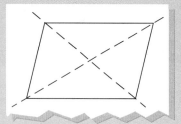

Money

1 Andrew bought 10 packets of sweets for £5.

How much did each packet cost?

10 ÷ 5 = 2
so each packet cost £2

2 Share £5.20 between four people.

5.20 ÷ 4 = 1.3

so each person receives one pound and three pence

3 Bars of chocolate cost £0.65 each.

What is the cost of 8 chocolate bars?

0.65 × 8 = 5.2
so the chocolate bars cost £5.2

㉕ Simple substitution

This will help you work out the value of expressions in algebra.

A Review

A1 Without using a calculator, work out each of these.

(a) $3 \times (6 - 1)$ (b) $(2 + 3) \times 10$ (c) $4 + 3 \times 4$

(d) $10 - (3 + 4)$ (e) $(4 - 1) \times 4$ (f) $8 - 6 \div 2$

A2 Handy Hire hire out tools.
For a ladder, they charge £2 per day, and an extra £5.

(a) How much will they charge for hiring a ladder for 6 days?

(b) How much will they charge for 10 days?

(c) Which of these rules is correct for hiring a ladder?
(c stands for the charge in £, n stands for the number of days' hire.)

$c = 5 \times n + 2$ $c = n + 2 + 5$ $c = 2 \times n + 5$ $c = 2 + 5$

A3 Here are some other things you can hire.
The charge is shown for each one.

Match each charge with one of the rules.
(There is one rule which does not match.)

Rules
$c = 2n + 3$ 2n means 2 × n
$c = 4n + 3$
$c = 3n + 4$
$c = 2n + 4$
$c = 3n + 2$

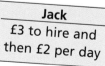

Jack
£3 to hire and
then £2 per day

Drill
£3 per day plus
an extra £4

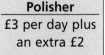

Polisher
£3 per day plus
an extra £2

Heater
£4 per day plus
an extra £3

A4 This is the hiring rule for a diamond cutter.

$$p = 5h + 6$$

p is the charge in £. h is the number of days hire.

(a) What is the charge for hiring the cutter for 3 days?

(b) How much does 6 days hire cost?

(c) Copy and complete this hire card for the cutter.

> **Cutter**
> £... per day plus
> an extra £...

A5

This is the rule for a floor sander.

$$p = 3h + 5$$

(a) What is p when h is 10?

(b) What is p when h is 20?

(c) What is the charge for hiring the sander for 4 days?

(d) If you have £20, what is the longest you could hire the sander for?

A6 Here are some more hire rules.
Work out what p is for each one when h is 6.

> $\frac{h}{2}$ means $h \div 2$

(a) $p = 4h$ (b) $p = h + 11$ (c) $p = \frac{h}{2}$

(d) $p = h + 3$ (e) $p = \frac{h}{3}$ (f) $p = 10h$

A7 For the rule $m = f + 12$, work out m when

(a) $f = 8$ (b) $f = 10$ (c) $f = 2$ (d) $f = 0$

A8 Use the rule $g = 2h + 8$ to work out g when

(a) $h = 10$ (b) $h = 5$ (c) $h = 100$ (d) $h = 4$

A9 For the rule $p = \frac{k}{2} + 1$, work out p when

(a) $k = 10$ (b) $k = 20$ (c) $k = 8$ (d) $k = 100$

A10 Which of these rules gives the biggest value of s when $t = 6$?

A $s = 4t + 10$ **B** $s = 6t - 5$ **C** $s = \frac{t}{2} + 20$ **D** $s = \frac{t}{3} + 30$

A11 Which of these rules gives the smallest value of s when $t = 6$?

A $s = t + 5$ **B** $s = 2t - 3$ **C** $s = \frac{t}{2} + 2$ **D** $s = \frac{t}{3}$

B The right order

When an expression has brackets, you work out what is in the brackets first.

To work out the value of $4(f-6)$ when $f = 11$.

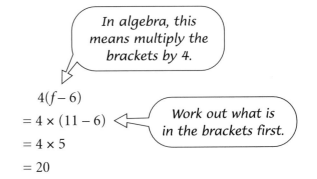

$4(f-6)$

$= 4 \times (11 - 6)$

$= 4 \times 5$

$= 20$

In algebra, this means multiply the brackets by 4.

Work out what is in the brackets first.

Where there are no brackets, multiply or divide **before** you add or subtract.

To work out the value of $4f - 6$ when $f = 11$.

$4f - 6$

$= 4 \times 11 - 6$

$= 44 - 6$

$= 38$

No brackets - so do the multiplication first.

B1 Copy and complete the working out for these expressions.

Work these out.

(a) $4(g - 1)$ when $g = 6$

$4(g - 1)$
$= 4 \times (6 - 1)$
$= 4 \times$ ▓
$=$ ▓

(b) $2(h + 3)$ when $h = 4$

$2(h + 3)$
$= 2 \times (4$ ▓
$=$ ▓
$=$ ▓

(c) $2(b + 5)$ when $b = 1$

$2(b + 5)$
$= 2 \times ($ ▓ $+ 5)$
$= 2 \times$ ▓
$=$ ▓

(d) $4(m - 2)$ when $m = 5$

$4(m - 2)$
▓
$=$ ▓
$=$ ▓

B2 Work out the value of each of these expressions when $a = 5$.

(a) $3(a - 2)$ (b) $4(a + 1)$ (c) $3a - 12$

(d) $5(a - 1)$ (e) $2a - 3$ (f) $6(a - 4)$

A B C D E F G H I J K L M N O P Q R S T U V W X Y Z

1 2 3 4 5 6 7 8 9 10 11 12 13 14 15 16 17 18 19 20 21 22 23 24 25 26

Work out the value of the clues for the words below.
Each clue gives one of the letters of the alphabet above.

For example, when $j = 2$, the clue $3j$ is 6, which would stand for **F**.

When you have finished, you will get a sentence.

1st word (two letters) $2u + 1$ when $u = 4$ and $4(p - 1)$ when $p = 6$

2nd word (two letters) $3(y + 1)$ when $y = 2$ and $4r - 1$ when $r = 5$

3rd word (six letters) $4f - 1$ when $f = 5$; $5(u - 2)$ when $u = 5$; $3(g + 1)$ when $g = 3$;
$5b + 2$ when $b = 4$; $2k - 3$ when $k = 4$; $4(s - 7)$ when $s = 8$

B4 Roland's River Runs organise parties on boats.
They use rules to work out what they need.

To work out how many paper plates they need
they add 5 to the number of people and then multiply by 2.

So for 15 people, they need $(15 + 5) \times 2$ plates, that is 40 plates.

(a) How many plates do they need for 25 people?

(b) How many plates do they need for 45 people?

(c) Which of these rules tells them how many plates to take?
n stands for the number of people, and
p stands for the number of plates.

A $p = 5n + 2$ **B** $p = 2(n + 5)$ **C** $p = 5(n + 2)$ **D** $p = 2n + 5$

B5 To work out the number of drinking straws they take,
multiply the number of people by 5, and then take off 10.

(a) How many straws do they need for 20 people?

(b) How many straws do 32 people need?

(c) Which of these rules tells them how many straws
to take for n people?
s stands for the number of straws.

A $s = 5(n - 10)$ **B** $s = 10(n - 5)$ **C** $s = 10n - 5$ **D** $s = 5n - 10$

B6 This is the rule for the number of glasses they need.

Add 15 to the number of people and then multiply by 3.

Which of these rules is correct for the number of glasses?
g stands for the number of glasses, n for the number of people.

A $g = 3(n + 15)$ **B** $g = 15(n + 3)$ **C** $g = 3n + 15$ **D** $g = 15n + 3$

B7 Here are some other rules they use.
Work out how many of each item they need
when there are 100 people.

$b = 2(n - 20)$

b is the number of balloons.

$t = \dfrac{n}{2} + 6$

t is the number of tables.

$p = 3(n + 50)$

p is the number of party poppers.

B8 Use the rule $h = 4(k + 12)$ to work out h when

  (a) $k = 5$ (b) $k = 0$ (c) $k = 3$ (d) $k = 12$

B9 The rule $t = 12 - s$ means 'take s away from 12'.

Work out t when

  (a) $s = 2$ (b) $s = 5$ (c) $s = 10$ (d) $s = 0$

*B10 We write 10^2 to mean 10×10.
In the same way, we write n^2 to mean $n \times n$.
So if $n = 6$, $n^2 = 6 \times 6 = 36$.

Work out n^2 when

  (a) $n = 3$ (b) $n = 5$ (c) $n = 1$ (d) $n = 4$

*B11 Work out

  (a) m^2 when $m = 5$ (b) $m^2 + 2$ when $m = 5$ (c) $m^2 - 20$ when $m = 5$

  (d) h^2 when $h = 4$ (e) $h^2 + 3$ when $h = 4$ (f) $d^2 - 10$ when $d = 6$

Grand Prix

A game for 2, 3 or 4 players.

- You need sheet 244 and a dice.
- You have to get round the board twice.
- First past FINISH the second time is the winner.

C Solving equations

$6 + \blacktriangledown = 10$

$8 - \text{✳} = 6$

$10 - x = 7$

$2b = 10$

$15 \div \blacksquare = 5$

$\bullet \times 3 = 12$

$2 + n = 8$

$\dfrac{a}{2} = 10$

C1 Solve each of these puzzles.

(a) $7 + \blacktriangledown = 10$ (b) $\text{✳} + 1 = 5$ (c) $\blacktriangle - 2 = 6$

(d) $2 \times \blacksquare = 12$ (e) $\blacklozenge \div 3 = 4$ (f) $\square \times 4 = 20$

C2 Find what each letter stands for in these equations.

(a) $n + 3 = 8$ (b) $2 + j = 12$ (c) $f - 1 = 7$

(d) $12 - t = 7$ (e) $100 - h = 98$ (f) $g + 10 = 15$

C3 Solve each of these equations.

(a) $2n = 10$ (b) $4b = 20$ (c) $3c = 21$

(d) $\dfrac{n}{2} = 5$ (e) $\dfrac{a}{3} = 4$ (f) $\dfrac{m}{2} = 6$

C4 Solve these equations.

(a) $20 - n = 16$ (b) $12 + f = 24$ (c) $12f = 24$

(d) $\dfrac{n}{5} = 2$ (e) $\dfrac{b}{2} = 2\frac{1}{2}$ (f) $28 - y = 6$

C5 Chloe and Tom are playing 'Snap' with equation cards.
Work out which pairs of cards have the same answer.

A $n + 3 = 10$

B $\dfrac{n}{2} = 3$

C $6n = 30$

D $n - 2 = 5$

E $10 - n = 5$

F $5n = 30$

*C6 Solve these equations.

(a) $2z + 3 = 17$ (b) $3w - 6 = 18$ (c) $4m + 5 = 29$

D Words to symbols

Sue has some coins in a bag.

Let **n** stand for the number of coins that Sue has.

Jaz has twice as many coins as Sue.

Then Jaz has **2n** coins …

Sandy has 10 more coins than Jaz.

… and Sandy has **2n + 10** coins.

D1

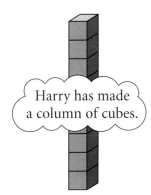

Harry has made a column of cubes.

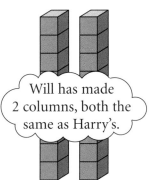

Will has made 2 columns, both the same as Harry's.

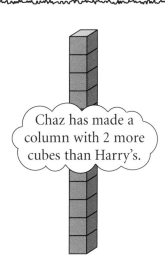

Chaz has made a column with 2 more cubes than Harry's.

Let *n* stand for the number of cubes in Harry's column.

(a) Which of the expressions below tells you the number of cubes Will has?

(b) Which of the expressions tells you the number of cubes Chaz has?

More than one of the expressions is correct in (a) and (b).

n^2 $n + n$ $2 + 2n$ $2n$ $n + 2$ $2n + 2$ $2 + n$

137

D2

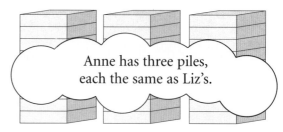

Let n stand for the number of tiles that Liz has.

(a) Which one of these expressions tells you the number of tiles Anne has?

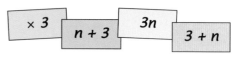

(b) Mary has three fewer tiles than Liz. Write an expression that tells you how many tiles Mary has.

(c) Sarah has three fewer tiles than Anne. Write an expression that tells you how many tiles Sarah has.

D3 Jake has some pancakes.   Kevin has four times as many pancakes as Jake.

Let p stand for the number of pancakes that Jake has.

(a) Write an expression for the number of pancakes that Kevin has.

(b) Diane has twice as many as Jake, and another 12.

Write an expression for the number of pancakes that Diane has.

D4 Annie has some dog biscuits.
Let n stand for the number of dog biscuits that Annie has.

(a) Bekky has 3 times as many biscuits as Annie. Write an expression for the number of biscuits Bekky has.

(b) Catch has 20 biscuits fewer than Annie. Write an expression for the number of biscuits Catch has.

What progress have you made?

Statement	Evidence

I can work out the value of simple expressions.

1 In the rule $s = 3t - 6$, work out s when
   (a) $t = 4$ (b) $t = 10$
   (c) $t = 3$ (d) $t = 2$

2 In the rule $w = \dfrac{d}{2} + 3$, work out w when
   (a) $d = 4$ (b) $d = 0$
   (c) $d = 12$ (d) $d = 1$

I can work out expressions with brackets.

3 Work out each of these expressions when $g = 6$.
   (a) $5(g - 4)$ (b) $4(g + 1)$
   (c) $2(g + 4)$ (d) $3(g - 1)$

I can write an expression for a situation in pictures.

4

Dan has some cubes. | Then he gets 3 more.

Let n stand for the number of cubes Dan had at the start.

Write an expression for the number of cubes he has after getting three more.

I can write an expression for a situation in words.

5 Sharon has some stamps.
Henry has three times as many stamps as Sharon.
James has 50 fewer stamps than Sharon.

Let s stand for the number of stamps that Sharon has.

   (a) Write an expression for the number of stamps that Henry has.

   (b) Write an expression for the number of stamps that James has.

26 Multiplication

This work will help you
- ◆ multiply together numbers like 23 and 34 without using a calculator
- ◆ multiply together numbers like 42 and 236 without using a calculator

A Review

A1 Find four matching pairs of multiplications.
Which is the odd one out?

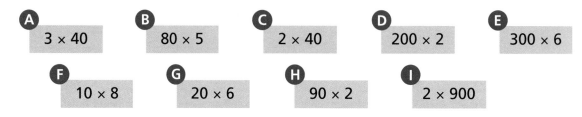

A   3 × 40   B   80 × 5   C   2 × 40   D   200 × 2   E   300 × 6

F   10 × 8   G   20 × 6   H   90 × 2   I   2 × 900

A2 Which of these is **not** equal to 180?

6 × 30   60 × 3   9 × 20   40 × 3   2 × 90

A3 Write down which of these multiplications give 1200.

4 × 30   200 × 6   12 × 10   3 × 400   6 × 20

A4 Work these out.
(a) 6 × 40   (b) 50 × 3   (c) 4 × 20   (d) 60 × 9   (e) 6 × 50
(f) 2 × 300   (g) 700 × 3   (h) 400 × 5   (i) 8 × 600   (j) 500 × 8

A5 A coach holds 40 passengers. It is full of holiday makers.
(a) Each holiday maker pays £6 for a seat on the coach.
How much money is this in total?
(b) How many passengers will five of these coaches be able to carry?

A6 Andrea took part in a sponsored swim.
She was sponsored £9 for each length.
She completed 30 lengths.

How much sponsor money should she collect?

A7 Copy and complete these multiplications.
(a) ? × 30 = 90   (b) 20 × ? = 80   (c) 400 × ? = 800
(d) ? × 3 = 900   (e) 7 × ? = 420   (f) 60 × ? = 300
(g) ? × 6 = 300   (h) ? × 9 = 4500   (i) 5 × ? = 4000

B More multiplication

30×20
$= 3 \times 10 \times 2 \times 10$
$=$

40×50
$= 4 \times 10 \times 5 \times 10$
$=$

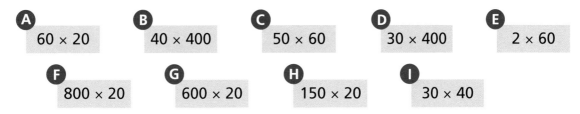

B1 Find four matching pairs of multiplications.
Which is the odd one out?

A 60×20 **B** 40×400 **C** 50×60 **D** 30×400 **E** 2×60

F 800×20 **G** 600×20 **H** 150×20 **I** 30×40

B2 Which of these are not equal to 4500?

90×50 50×900 30×150 900×50 70×50

B3 Write down which of these multiplications give 6000.

300×20 20×30 20×300 $20 \times 30 \times 10$

B4 Work these out.

(a) 20×40 (b) 30×70 (c) 20×50 (d) 90×90

(e) 50×60 (f) 400×20 (g) 60×300 (h) 50×500

(i) 80×50 (j) 200×70 (k) 90×60 (l) 30×900

B5 Copy and complete these multiplications.

(a) $? \times 30 = 2100$ (b) $40 \times ? = 2800$ (c) $? \times 80 = 4000$

(d) $60 \times ? = 300$ (e) $90 \times ? = 2700$ (f) $? \times 20 = 14\,000$

B6 Write down five multiplications that have 2400 as the answer.
Here is one to start you off: $2 \times 1200 = 2400$

B7 A football pitch measures 90 metres by 80 metres.
What is its area?

B8 How much does it cost for 40 cans of drink that cost 40p each?

B9 Find the area of these rectangles.

(a)

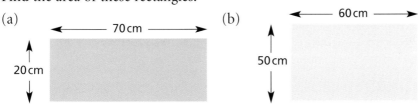

(b)

C Area and multiplication

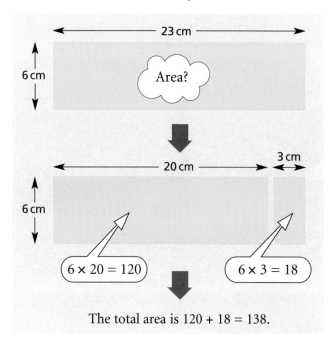

The total area is 120 + 18 = 138.

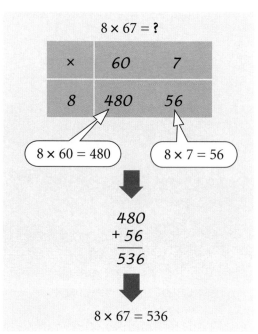

$8 \times 67 = 536$

C1 This large rectangle has been split into two smaller rectangles.

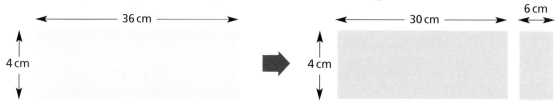

   (a) Find the area of the two smaller rectangles.

   (b) Write down the area of the large rectangle

C2 Work out the area of these rectangles by splitting them into two smaller ones.

   (a)

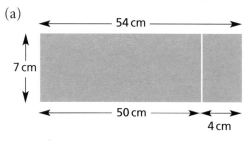

   (b)

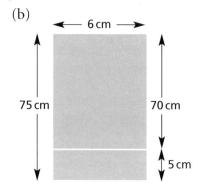

C3 Find the areas of these rectangles by splitting each one into smaller rectangles. Make rough sketches to show how you split each one.

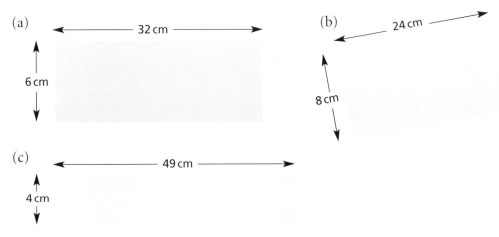

(a) 32 cm, 6 cm

(b) 24 cm, 8 cm

(c) 49 cm, 4 cm

C4 Copy and complete this table to work out 7 × 28.

×	20	8
7		

C5 Copy and complete this table to work out 9 × 14.

×	10	...
9		

C6 Draw tables to work these out.

(a) 5 × 19   (b) 7 × 72   (c) 46 × 6   (d) 83 × 5

C7 Work out these multiplications without a calculator.

(a) 6 × 23   (b) 7 × 56   (c) 4 × 63   (d) 64 × 5

C8 A disco ticket cost £9 per person.
How much will 81 tickets cost?

C9 A woman spends £4 a week on the football pools.
How much is this over a 33 week football season?

C10 How many days are there in 48 weeks?

C11 Sam uses 42 litres of petrol a week.
How much is this over eight weeks?

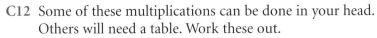

C12 Some of these multiplications can be done in your head.
Others will need a table. Work these out.

(a) 7 × 26   (b) 60 × 3   (c) 5 × 70   (d) 6 × 70

D Tables to multiply

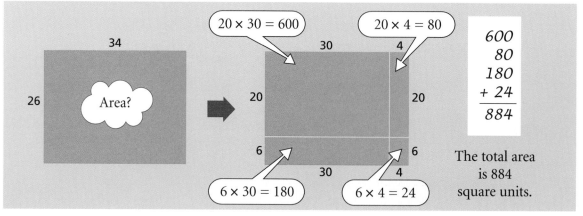

×	30	4
20	600	80
6	180	24

$26 \times 34 = ?$

$$\begin{array}{r} 600 \\ 80 \\ 180 \\ + \ 24 \\ \hline \end{array}$$

$26 \times 34 = 884$

D1 (a) Copy and complete this table.

(b) Use the table to work out 47×25.

(c) Make a table for 46×26 and work it out.

×	20	5
40		
7		

D2 Work out 37×24 by filling in a table like this.

×	20	4
30		
7		

D3 Make tables for these multiplications and use them to work out the answers.

(a) 19×16 (b) 13×27 (c) 21×34 (d) 45×25

(e) 15×15 (f) 23×91 (g) 96×41 (h) 62×87

D4 In a school hall the seats are laid out in 21 equal rows.
There are 16 seats in each row.

How many seats are there altogether?

D5 Andrea took part in a sponsored swim. She was sponsored £23 for each length.
She completed 37 lengths. How much sponsor money should she collect?

D6 Find the area of a rectangle that measures 56 cm by 22 cm.

D7 This table can be used to work out 45 × 342.

Copy and complete this working to find the answer to 45 × 342.

×	300	40	2
40	12 000	1600	80
5	1500		

```
        12000
         1600
           80
         1500
          ...
     +    ...
45 × 342 =  ...
```

D8 Make tables to work these out.

(a) 146 × 23   (b) 35 × 268   (c) 108 × 29   (d) 26 × 254

(e) 324 × 42   (f) 15 × 658   (g) 452 × 54   (h) 62 × 324

D9 Work out these multiplications.
You will not need a table for all of them.

(a) 425 × 34   (b) 300 × 40   (c) 60 × 700   (d) 32 × 415

Four digits

- You have four digits, 1, 2, 3, 4

  How many different multiplications can you make of the form ☐☐ × ☐☐ ?

  $14 × 32$

  $12 × 34$ $21 × 34$

  Which gives the largest result?
  Which gives the smallest?

- Can you find the largest and smallest results for the digits 2, 3, 4, 5?

- Investigate for different sets of digits.
  Can you say how to get the largest and smallest results with any set of four digits?

- Investigate the smallest and largest multiplication you can make using 1, 2, 3, 4, 5.

What progress have you made?

Statement

Evidence

I can do multiplications like 30 × 40 and 60 × 500 without a calculator.

1 Work these out.

(a) 3 × 50   (b) 30 × 30

(c) 40 × 80   (d) 50 × 60

(e) 80 × 90   (f) 500 × 70

I can do multiplications like 34 × 472 without a calculator

2 Work these out.

(a) 15 × 26   (b) 27 × 41

(c) 39 × 29   (d) 78 × 92

(e) 34 × 263   (f) 431 × 52

27 Better or worse?

This is about comparing different sets of data.
The work will help you

♦ remember the difference between the mean, median and range

♦ use the mean and range to compare sets of data

A Mean, median and range

Play the game 'A mean roll' on sheet 245.

A1 Charlie counts the number of people in cars passing his house.
Here is his data, written out in order.

1, 1, 1, 1, 1, 1, 2, 2, 2, 3, 3, 4, 4, 4, 6

(a) Find the median number of people in a car.

(b) What is the range of the number of people in the car?

(c) Calculate the mean number of people in a car.

A2 Alison has a number of pepper plants in her greenhouse.
She has counted the number of peppers growing on each plant.

4, 10, 4, 5, 7, 5, 5, 6, 3, 6, 5, 6

(a) Find the median number of peppers on a plant.

(b) What is the range of the number of peppers on a plant?

(c) Calculate the mean number of peppers on a plant.

A3 David was carrying out a survey on healthy eating.
He asked some friends how many pieces of fruit they had eaten the previous day.

3, 2, 0, 0, 1, 1, 0, 2, 1, 2, 5, 2, 0, 2

(a) What was the median number of pieces of fruit eaten?

(b) Find the range of the number of pieces of fruit eaten.

(c) What was the mean number of pieces of fruit eaten?

A4 Sanjay records the number of hours of sunshine at 10 British seaside towns
on a day in August.

2.2   1.9   5.6   4.4   4.0   2.7   3.8   2.3   1.7   1.4

Find (a) the median   (b) the range   (c) the mean of the hours of sunshine

A5 Conor rolls five dice but hides the score on one of them.
The scores on the other four dice are   5, 3, 3, 4

  (a) He says that the range of his scores is 4.
  What was the score on the 5th dice?

  (b) What was the median of his scores?

  (c) What was the mean of his scores?

A6 Asher rolls five dice and also hides one of the scores.
The scores on the other four dice are   3, 6, 5, 5

  (a) Asher says that the mean of the scores is exactly 5.
  What is the score on Asher's 5th dice?

  (b) Use this to find the median and range of Asher's five dice scores.

Four lives A game for two players

  • The first player rolls five dice, and does not show them to the second player.

  • The first player tells the second what the mean, median and range of the scores is.

  • The second player then tries to guess what the scores on the hidden dice are.

  • If the second player cannot guess, they can ask to see the dice, one at a time.

  • The second player scores a point for each dice still hidden when they guess correctly.

  • The players swap over and play again.

B Decimal means

You need to work in groups for this activity.

How good are you at estimating? Estimate   • 10 cm   • 60°   • 1 minute   • 5 metres.

Work out the mean of each of your group's estimates.
Were they higher or lower than the measurements you were trying to estimate?

B1 Work out the mean of each of these sets of data.
Give your answers correct to one decimal place.

  (a) The weights of some boys in kg: 37 51 41 45 52 57

  (b) The heights of some girls in cm: 130 123 141 150 140 134 126

  (c) The distance some pupils travel to school in km:

   2.6 0.9 3.2 4.5 7.8 1.8 2.6 3.5 1.3

  (d) The lengths of the world's main commercial aircraft in metres:

   28.9   32.9   32.8   64.3   28.5   38.1   45.6   29.0

   34.1   50.4   44.8   27.4   44.4   45.2   28.5   25.6

147

B2 These tables show the points scored by the winning drivers in the Formula One World Championships in the 1980s and the 1990s.

Year	Winner	Points
1980	Jones	67
1981	Piquet	50
1982	Rosberg	44
1983	Piquet	59
1984	Lauda	72
1985	Prost	73
1986	Prost	72
1987	Piquet	73
1988	Senna	90
1989	Prost	76

Year	Winner	Points
1990	Senna	78
1991	Senna	96
1992	Mansell	108
1993	Prost	99
1994	Schumacher	92
1995	Schumacher	102
1996	Hill	97
1997	Villeneuve	81
1998	Hakkinen	100
1999	Hakkinen	76

(a) Calculate the mean number of points of the winner in the 1980s.

(b) Calculate the mean number of points of the winner in the 1990s.

(c) In which period, on average, were more points gained by the winner?

C Comparisons

For class discussion

A PE teacher has to select two girls to represent the school at the high-jump.
She looks at the last few jumps (in centimetres) of some pupils.

Emma	148	145	147	138	143
Shula	129	146	145	139	141	147
Karin	152	128	150	138	139
Bonnie	140	141	144	142
Gita	149	147	141	145	146

Which two pupils should the teacher choose to represent the school?

C1 Pauline is deciding where to go on holiday to get the warmest weather.
She looks at the temperatures in two resorts for the same two weeks in August.
Here are the highest daily temperatures in °C.

Bournemouth 19 17 19 20 21 20 22 17 22 23 23 21 23 21

Eastbourne   20 20 18 20 21 19 20 19 21 20 21 21 22 20

(a) Calculate the mean and range of the temperatures at Bournemouth.

(b) Calculate the mean and range of the temperatures at Eastbourne.

(c) Which resort had the highest temperatures on average?

(d) Which resort had the greatest variation in temperature?

C2 These two graphs show the number of hours of daily sunshine at two resorts during the same week in August.

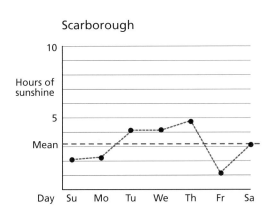

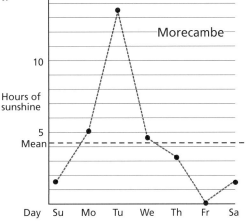

(a) From the graph, about what was the mean number of hours of sunshine in Scarborough?

(b) Which resort had more sunshine on average?

(c) Which resort had most number of days with more than 2 hours of sunshine?

(d) Which resort had the biggest range of hours of sunshine?

(e) Say, with reasons, which resort you would rather have stayed at.

C3 This chart shows how many papers each house in Avenue Road had delivered one day.

(a) How many houses had 1 paper delivered?

(b) What was the most common (the modal) number of papers delivered?

(c) How many houses are there in Avenue Road altogether?

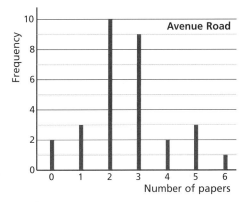

C4 This chart shows the number of papers delivered to each house in Orchard Crescent the same day.

(a) How many houses had no paper delivered?

(b) How many had 3 or more delivered?

(c) What was the modal number of papers delivered in Orchard Crescent?

C5 Just by looking at the charts, in which road do you think the houses had more papers delivered?

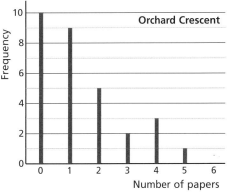

C6 These two charts show the number of letters delivered to the houses in Rose Crescent and Paradise Drive one day.

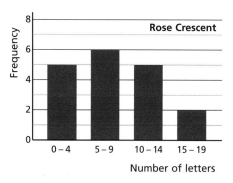 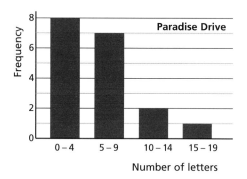

(a) (i) In Rose Crescent, how many houses had 4 or less letters delivered?

(ii) How many houses had 4 or less letters in Paradise Drive?

(b) In Rose Crescent the most common number of letters delivered was between 5 and 9. We call this the **modal class.**
What was the modal class for the letters delivered in Paradise Drive?

(c) In which road do you think more letters were delivered?

What progress have you made?

Statement	Evidence
I can find the median, mean and range of a set of data using decimals.	1 The price in pence per litre of petrol in EU countries in August 1999 was 73.6, 72.9, 68.8, 66.9, 65.9, 65.6, 64.9, 62.1 60.6, 58.6, 58.4, 55.6, 49.1, 48.8, 48.4 (a) Find the median and range of these prices. (b) Calculate the mean of these prices.
I can compare two sets of data using the mean and range.	2 The price of two brands of margarine in different shops were (in pence) **Sungold** 79 82 67 81 75 78 85 70 **Highlife** 74 76 73 79 69 77 76 (a) Find the mean and range of the price of Sungold. (b) Find the mean and range of the price of Highlife. (c) Which margarine was cheaper on average? (d) Which margarine varied the least in price?

Review 4

1

A	B	C	D	E	F	H	K	O	S	T	W
0.02	0.05	0.06	0.2	0.5	0.6	2	5	6	20	50	60

Work out the answers to each of these (do not use a calculator).
Use the code above to find a letter for each answer.
Rearrange each set of letters to make the name of a tree.

(a) 0.2 × 10   2 ÷ 100   0.2 × 100

(b) 0.5 × 10   0.2 ÷ 10   0.06 × 100

(c) 5 ÷ 10   0.6 ÷ 10   5 ÷ 100   0.02 × 100   50 ÷ 100

2 Write these animals in order of weight, lightest first.

| Rat | Puppy | Kitten | Guinea pig | Rabbit |
| 0.3 kg | 220 g | 90 g | 0.5 kg | 1.1 kg |

3 You can use this graph to convert
between feet and metres.

(a) What is 70 metres in feet?

(b) What is 150 feet in metres?

4 The Comet was the first passenger jet aircraft.
It was 93 feet long, had a wingspan of 115 feet
and carried 36 passengers.

The Airbus 380 has a length of 54 metres
and a wingspan of 45 metres.
It will carry 656 passengers.

(a) Which is longer, the Airbus 380 or the Comet?
By how many metres?

(b) Which has the bigger wingspan of the
two aircraft? By how many metres?

(c) How many more passengers will the
Airbus 380 carry compared to the Comet?

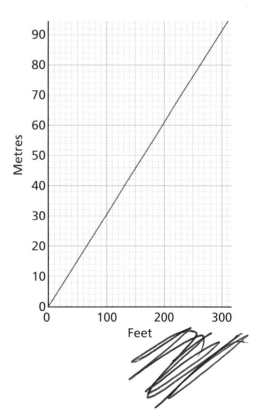

5 Spot the mistake in each of these. Correct the mistake
  and say why you think it was made.

(a) I take $\frac{1}{4}$ kg from a
1 kg bag of flour.
How much is left?

$1000 - \frac{1}{4}$
$= 999\frac{3}{4}$ grams.

(b) Plot A at (⁻1, 2).

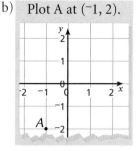

(c) Shade squares so the
shape has rotation
symmetry order 2.

(d) Phil has £10.
He spends 2p.
How much does
he have left?

$10 - 2 = 8$.
He has £8 left.

6 Work out the value of each expression when $a = 8$.

(a) $a + 2$ (b) $a - 5$ (c) $3a$ (d) $3a - 4$ (e) $2(a - 1)$

(f) $3(a + 1)$ (g) $4a + 2$ (h) $\frac{a}{4} - 1$ (i) $20 - a$ (j) a^2

7 Solve each of these equations.

(a) $x + 4 = 6$ (b) $2x = 8$ (c) $x - 4 = 6$ (d) $\frac{x}{4} = 2$

8 Without using a calculator, work these out.

(a) 20×30 (b) 400×30 (c) 40×50 (d) 500×80

9 Copy and complete this table
to work out 23×34.

×	30	4
20		
3		

10 Without a calculator, work out 24×325.
Show all your working clearly.

11 Afzana counted the number of birds that were
at her bird feeder every five minutes for an hour.

2 1 3 5 0 2 6 0 2 4 5 6

(a) What was the median number of birds?

(b) What was the mean number of birds?

(c) What was the range of the number of birds?

12 Shakib counted the number of birds on
his bird feeder.

1 3 5 2 3 2 2 3 2 3 3 1

(a) Find the mean and range of the number
of birds on Shakib's feeder.

(b) Which feeder had the bigger mean number of birds,
Afzana's or Shakib's?

(c) Which feeder had the bigger range?